1 MONTH OF
FREE
READING

at

www.ForgottenBooks.com

By purchasing this book you are eligible for one month membership to ForgottenBooks.com, giving you unlimited access to our entire collection of over 1,000,000 titles via our web site and mobile apps.

To claim your free month visit:

www.forgottenbooks.com/free311126

ISBN 978-0-483-79358-3
PIBN 10311126

By **FRANK STEPHENS**

Illustrated by **W. J. Fenn**
from studies in the field

Published by
The **West Coast Publishing Co.**
San Diego, California
1906

Contents:

Introduction

The area treated of in this volume is strictly California and that part of the Pacific Ocean properly belonging to California. All the mammals described are known to have been found within the State or within sight of its shores. The number of species and subspecies proves to be very large. This is accounted for partly by the large extent and great latitudinal length of the State, but more by the very great variety of climate within the State, greater than occurs in any other State of the Union; grading all the way from the subtropical region of the Colorado Valley and Desert to the arctic climate of the eternal snows on the summits of the Sierra Nevada.

No general work covering the mammals of this State has been published since 1857, when Baird's Vol. VIII of the Pacific Railroad Reports was issued. This did not contain the marine mammals, the bats nor man. A great advance in our knowledge of the land mammals has occurred within the last fifteen years, and some of the material obtained in this time has not yet been critically examined by systematic experts, hence we may expect further additions to the present known species, as well as more or less revision of the nomenclature.

The distribution of species herein given has been checked in the majority of species from personal observation. I have done no field work in the northwestern part of California and but little in the northeastern part. I have had exceptionally good opportunities for observation for many years in the southern part of the State and I believe the statements of distribution for this part will bear close inspection.

There is no "royal road to knowledge." This saying is true of all the natural sciences and mammalogy is no exception. The beginner will find it difficult to get a start, but when one becomes a little familiar with the general characters of the larger groups it is a comparatively easy matter to trace out a species and learn its name, which should be but a preliminary step to further study of the species, and not the end as is but too often the case. Of necessity the division into orders, families and genera are made on technical characters, and it is better for the student to master these and begin aright. I have used technical terms as little as practicable. Their moderate use admits of much greater conciseness of description. To avoid the use of technical terms would necessitate the use of cumbersome expressions that would greatly increase the size of this volume. For the explanation of the technical terms refer to the glossary, in front of the index.

The full description of a mammal includes not only the characters given under the specific name, but also the characters previously given under its genus, family and order; to add these each time in the specific description would be confusing as well as cumbersome. After becoming a little familiar with the subject it will not be necessary to refer to these higher characters each time.

A departure from recognized usage in the use of names of authorities is made for the sake of simplicity. The authority for a specific or subspecific name is given without reference to generic changes made later. The words in parentheses after the technical name are intended to be a translation of the Latin or Greek name. This translation is sometimes a free one, to give the sense of the name intended.

The measurements used are —"length," the distance from the tip of the nose to the end of the skin of the tail, taken with the animal laid on its back on the scale; "tail vertebræ," taken with the dividers with one point set on the rump at the base of the tail, the tail being held at right angles to the body, the other point being placed at the end of the skin of the tail, "hind foot," the distance from the end of the longest claw to the upper edge of the heel, the true heel being used, which in many mammals is not the termination of the sole, but in such animals as the cat, deer or dog what is popularly, but wrongly, called the knee; "ear from crown," taken with one point of the dividers set on the skull on the inner (convex) side of the ear and the other at the tip of the ear.

In the dental formula "I" means incisor teeth; "C" canine; "P" premolar; "M" molar. The number means the number of that class of teeth in one side of the upper or lower jaw, respectively; the last number being the total of all teeth.

The standard used for the names of colors is Ridgways "Nomenclature of Colors."

The measurements are given in millimeters as being better adapted for the use of naturalists; they are practically duplicated in inches and hundreths, in parentheses, for the use of those students who have no metric scale. The following table for the conversion of inches into millimeters and *vice versa* may be useful.

Inch.	Mm.	Millimeter.	Inch.
1	25.39	1	.0393
2	50.78	2	.0787
3	76.18	3	.1181
4	101.57	4	.1574
5	126.97	5	.1968
6	152.36	6	.2362
7	177.76	7	.2755
8	203.15	8	.3150
9	228.55	9	.3543
10	253.94	10	.3937

This book is little more than a mere dry skeleton; if it aids the student in finding out for himself or herself some portions of the life histories of our mammals I shall be pleased. I have labored under the disadvantage of being out of reach of good reference libraries. Nearly all the workers in this field have sent me copies of their papers as soon as published; without this help this volume would have been of little real value. I have made free use of all such papers, but to save space I have seldom given the authority for statements made. In many cases the facts have been condensed from several authorities into the briefest possible statement. I would like to acknowledge by name the aid received through this and other sources but the number is very great and to mention but part would be unfair to the remainder. The list would include the name of practically every one who has done field work among the mammals of California, or has written on material coming from this State; hence this volume is really a compilation of all the work done on Californian Mammals, and each author or collector may consider that he has a share in whatever merit it may possess.

California Mammals
Class **Mammalia.** **Mammals**

Young born alive and nourished by milk secreted in mammæ; lungs and heart contained in a thorax separated from the abdominal viscera by a diaphragm; heart four chambered; circulation complete; blood warm, with red non-nucleated corpuscles; body usually covered with hairs; mouth usually furnished with teeth; never more than two pairs of limbs, both pairs always present except in some aquatic species.

Subclass **Monodelphia.**

Anterior cerebral commissure small; corpus callosum large; episternum wanting; coracoid very feebly developed, not connected with a sternum; urogenital and intestinal openings not combined; a placenta; young well developed when born.

Order **Cete.**
WHALES, DOLPHINS, PORPOISES, ETC.

Fore limbs fin-like, without distinct fingers and without nails; hind limbs absent; pelvis rudimentary; no clavicles; tail widened horizontally; neck short, the vertebræ more or less fused; nostrils opening on top of the head as spiracles; eyes small; no external ear; skin hairless; habitat marine.

Cetaceans are mammals that are fishlike in form and adapted to life in oceans, seas and large rivers. Like all mammals cetaceans breathe by means of lungs and suckle their young, which are born well developed.

The only book containing full and accurate accounts of the habits of our species is the "Marine Mammals of the Northwestern Coast of North America," by Captain C. M. Scammon, pub-

lished in 1874.   As it is now very scarce and inaccessible to the
general public I shall give considerable space to extracts from it.
I have very little direct personal knowledge of this order.

Suborder **Mystacete.**   (Mustache—whale.)

No teeth present after birth; upper jaws furnished with
plates of baleen (whalebone;) rami of lower jaw connected by
fibrous tissue and not by a suture; olfactory organ developed;
spiracle double.

Family **Balænidæ.**   Whalebone Whales.

Lower jaw very thick and deep; cleft of mouth curved; skull
symmetrical.

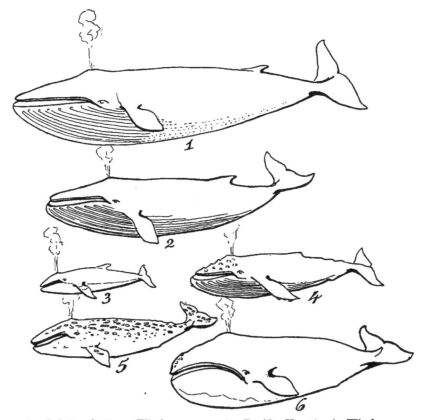

1. Sulphur-bottom Whale.   4. Pacific Humpback Whale.
2. Oregon Finback Whale.   5. California Gray Whale.
3. Sharp-headed Finner Whale.   6. Pacific Right Whale.

About twenty species, divided among eight genera. The species are marine, usually pelagic. They are found in all seas but are least common in tropical seas. Owing to the difficulty of preserving the parts of such huge animals but little material has been examined by competent naturalists, and therefore but little is accurately known of their relationships.

The food of Whalebone Whales is zoophytes, molluscs, crustaceans and small fish. When a quantity of these are taken in the mouth the water is strained out through fringed baleen; the mouth being partially closed. The throat is comparatively small, the food being animals of quite small sizes.

Genus **Balæna** Linn. (Whale.)

No dorsal fin; pectoral fin short, broad and enclosing the bones of all five fingers; head very large; baleen very long, narrow, black; cervical vertebræ united; skin of throat not furrowed.

Balæna japonica Gray. (Of Japan.)
PACIFIC RIGHT WHALE.

Large; head large in proportion; color black, occasionally spotted with white. Length about sixty feet.

Arctic and North Pacific Oceans.

"The color of the Right Whale is generally black, yet there are many individuals with more or less white about their throat and pectorals ,and sometimes they are pied all over. The average length may be calculated at sixty feet—it rarely attains to seventy feet—and the two sexes vary little in size. The head is very nearly one third the length of the whole animal, and the upper intermediate portion, or that part between the spiracles and 'bonnet.' has not that eve nspherical form, or the smooth and glossy surface present with the Bowhead, but is more or less ridgy crosswise. Both lips and head have wart-like bunches moderately developed, and in some cases the upper surface of the head and fins are infested with parasitical crustaceans.

"The tongue yields oil like the *mysticetus,* but its baleen is shorter and of a coarser and less flexible nature. The average product of oil of the *Balæna japonica* may be set down as one hundred and thirty barrels. The amount of bone ranges from one thousand to fifteen hundred pounds.

"In former times the Right Whales were found on the coast of Oregon, and occasionally in large numbers. The few frequenting the coast of California are supposed to be merely stragglers for their northern haunts." (Scammon.)

Genus **Rhachianectes** Cope. (Spine—Swimming.)

No dorsal fin; pectoral fin narrow, enclosing the bones of but four fingers; head comparatively small, baleen short, coarse; cervical vertebræ free; skin of throat with two longitudinal furrows.

Rhachianectes glaucus Cope. (Whitish blue.)
CALIFORNIA GRAY WHALE.

Size medium; color varying from light mottled gray to nearly black.

Length of female about forty feet; male is smaller.

Pacific coast of North America.

"The California Gray Whale is unlike the species of *Balæna* in its colors, being of a mottled gray, very light on some individuals, while others, both male and female, are nearly black. Under the throat are two longitudinal folds, which are about fifteen inches apart and six feet in length. The eye is situated about five inches above and six inches behind the angle of the mouth. The ear, which appears externally like a mere slit in the skin, two and a half inches in length, is about eighteen inches behind the eyes and a little above it. The length of the female is forty to forty-four feet; its greatest cicumference twenty-eight to thirty feet; its flukes thirty inches in depth and ten to twelve feet broad; its pectorals about six and one-half feet in length and three feet in width, tapering from near the middle toward the ends, which are quite pointed. It has no dorsal fin; usually the limbs of the animal vary but little in proportion to its size. The male may average thirty-five feet in length, but varies more in size than the female.

"The blubber is six to ten inches in thickness. The average yield of oil is twenty barrels. The baleen, of which the longest portion is fourteen to sixteen inches, is of a light brown color, the grain very coarse.

"The California Gray Whale is only found in north latitudes, and its migrations have never been known to extend lower than 20 degrees north. It frequents the coast of California from November to May. During these months the cows enter the lagoons on the lower coast to bring forth their young, while the males remain outside along the seashore. The time of gestation is about a year. Occasionally a male is seen in the lagoons with the cows toward the end of the season, and soon after both male and female, with their young, will be seen working their way northward, following the shore so near that they often pass through the kelp near the beach. It is seldom that they are seen far out at sea. This habit of resorting to shoal bays is one in which they differ strikingly from other whales.

"In summer they congregate in the Arctic Ocean and Okhotsk Sea. It has been said that this species of Whale has been found off of the coast of China and about the shores of Formosa, but the report needs confirmation. In October and November the California Gray Whales appear off the coast of Oregon and Upper California, on their way back to their tropical haunts, making a quick, low spout at long intervals; showing themselves very little until they reach the smooth lagoons of the lower coast, where, if not disturbed, they gather in large numbers, passing into and out of the estuaries, or slowly raising their collosal forms midway above the surface, falling over on their sides as if by accident and dashing the water into foam and spray about them. At times, in calm weather, they are seen lying in the water quite motionless, keeping one position an hour or more.

At such times the sea gulls and cormorants alight on the huge beasts.

"From what data we have been able to obtain the whole number of California Gray Whales which have been captured or destroyed since the bay whaling commenced, in 1846, would not exceed 10,800, and the number which now periodically visit the coast does not exceed 8,000 or 10,000." (This appears to have been written in 1872.)

"Many of the marked habits of the California Gray Whale are widely different from those of any other species of *Balæna*. It makes regular migrations from the hot southern latitudes to beyond the Arctic Circle; and in the passage between the extremes of climate it follows the general trend of an irregular coast so near that it is exposed to the attacks of the savage tribes inhabiting the seashore, who pass much of their time in the canoe, and consider the capture of this singular wanderer a feat worthy of the highest distinction. As it approaches the waters of the torrid zone, it presents an opportunity to the civilized whalemen—at sea, along shore, and in the lagoons—to practice their different modes of strategy, thus hastening the time of its utter annihilation. This species of whale manifests the greatest affection for its young, and seeks the sheltered estuaries lying under a tropical sun as if to warm its offspring and promote its comfort, until grown to a size Nature demands for its first northern visit." (Scammon.)

Genus **Megaptera** GRAY. (Large—fin.)
Dorsal fin present, low or "hump" like; pectoral fin very long and narrow; head of moderate size; throat and belly with longitudinal furrows; baleen short; cervical vertebræ free; size large.

Megaptera nodosa versabilis COPE. (Knotted; capable of being turned.)
PACIFIC HUMPBACK WHALE.
Body short and thick; a "hump" of variable size and shape

situated similarly to the dorsal fin of other species; pectoral and caudal fins very large; color black, more or less mottled with white below.

Pacific Ocean.

"The Humpback is one of the species of rorquals that roam throughout every ocean, generally preferring to feed and perform its uncouth gambols near extensive coasts, or about the shores of islands, in all latitudes between the equator and the frozen oceans, both north and south. It is irregular in its movements, seldom going in a straight course for any distance, at one time moving about in numbers, scattered over the sea as far as the eye can discern from the masthead; at other times singly, seemingly as much at home as if surrounded by hundreds of its kind.

"Its shape, compared with the symmetrical forms of the Finback, California Gray and Sulphurbottom, is decidedly ugly, as it has a short, thick body, and frequently a diminutive 'small', with inordinately large pectorals and flukes. A protuberance, of variable size and shape in different individuals, placed on the back, about one fourth the length from the caudal fin, is called the hump. Another cartilaginous boss projects from the center fold immediately beneath the anterior point of the lower jaw, which, with the flukes, pectorals and throat of the creature, are often hung with pendant parasites (*Otion stimpsoni*), and on the males it is frequently studded with tubercles, as on the head. The under jaw extends forward considerably beyond the upper one. All these combined characteristics impress the observer with the idea of an animal of abnormal proportions. The top of the head is dotted with irregular rounded bunches, which rise about an inch above the surface, each covering about four square inches of space.

"Extreme length (of a male) 49 feet 7 inches; length of pectoral 13 feet 7 inches; breadth of pectoral 3 feet 2 inches; expansion of flukes 15 feet 7 inches; breadth of flukes 3 feet 4 inches; length of folds on belly 16 feet; thickness of blubber 5 to 10 inches; color of blubber yellowish white; yield of oil 40 barrels; number of folds on belly 26, averaging four to six inches in width. These folds, which extend from the anterior portion, of the throat over the belly, terminating a little behind the pectorals, are capable of great expansion and contraction, which enables the Humpback, as well as the other rorquals, to swell their maws when food is in abundance about them. It is proper to state that the skull and upper jaw bone of any ordinary sized animal would be about 15 feet long by 6 broad.

"The usual color of the Humpback is black above, a little lighter below, slightly marbled with white or gray; but sometimes the animal is of spotless white under the fins and about the abdomen. The posterior edge of the hump, in many animals, is tipped with pure white.

"The *Megaptera* varies more in the production of oil than all others of the rorquals. We have frequently seen individuals which yielded but 6 to 10 barrels of oil, and others as much as 75. Most of this variation may be attributed to age or sex.

"Like all other rorquals it has two spiracles, and when it respires the breath and vapor ejected through these apertures form the 'spout,' and rise in two separate colums, which, however, unite as they ascend and expand. When the enormous lungs of the animal are brought into full play the spout ascends twenty feet or more. When the whale is going to windward, the influence of the breeze is such that a low bushy spout is all that can be seen. The number of spouts to a 'rising' is exceedingly variable; sometimes the animal blows only once, at another time up to 15 or 20 times.

"Although the Humpback is found on every sea and ocean, our observations indicate that they resort periodically, and with some degree of regularity, to particular localities, where the females bring forth their young. It seems, moreover, that both sexes make a sort of general migration from the warmer to the colder latitudes, as the seasons change. They go north in the northern hemisphere as the summer approaches, and return south as winter sets in.

"In the Bay of Monterey, Upper California, the best season for Humpbacks is in the months of October and November; but some whales are taken during the period from April to December, including a part of both months. The great body of these whales, however, are observed working their way northward until September, when they begin to return southward; and the Bay being open to the north, many of the returning band follow its shores or visit its southern extremity, in search of food, which consists principally of small fish, or the lower orders of crustaceans. When the animals are feeding the whalers have a very favorable opportunity for their pursuit and capture.

"The Humpbacks are captured with a common hand-harpoon and lance, 'Greeners Harpoon Gun,' and the bomb-lance, by the whaleships crew; and as they are very liable to sink when dead, every exertion is made to get the harpoon in, with line attached, before the bomb gun is discharged. Then if the creature goes to the bottom, a buoy is attached to the end of the line, or a boat lies by it, until the decomposition of its flesh has generated sufficient gas to allow the animal to be drawn up. The length of time that elapses before this takes place of course depends much on the depth of the water and the solidity of the animals formation; some individuals remaining but a few hours on the bottom, while others will remain down two or three days at the same depth.

"The best points for Humpback whaling on the coast have been Magdalena, Ballenas and Monterey Bays; but since the acquisition of Alaska numerous places have been found in the bays and about the islands of that Territory, which doubtless in the future will become profitable whaling stations."

Genus **Balænoptera** LAPECEDE. (Whale—fin).

Dorsal fin small, curved; pectoral fin small, narrow; head flat; body slender, skin of throat with longitudinal furrows; baleen short and coarse; cervical vertebræ free.

Balænoptera physalus velifera Cope. (Sail bearing.)
OREGON FINBACK WHALE.

Large; color blackish above, white below.

North Pacific Ocean.

"A Finback picked up by Captain Poole of the bark Sarah Warren, of San Francisco, affords us the following memoranda; length 65 feet; thickness of blubber 7 to 9 inches; yield of oil 75 barrels; color of blubber a clear white; top of head quite as flat and straight as that of a Humpback; baleen, the longest, 2 feet 4 inches; the greatest width 13 inches; its color a light lead streaked with black, and its surface presents a ridgy appearance crosswise, length of fringe to bone 2 to 4 inches, and in size the fringe may be compared to a cambric needle.

"A *Balænoptera* which came ashore near the outer heads of the Golden Gate gave us the opportunity of obtaining the following rough measurements; length 60 feet; from nib-end to pectorals 15 feet; from nib-end to corner of mouth 12 feet; expansion of caudal fins 14 feet. Its side fins and flukes are in like proportion to the body as in the California Gray. Its throat and breast are marked with deep creases or folds, similar to the Humpback. Color of sides and back black or blackish brown (in some individuals a curved band of lighter shade marks its upper sides, between the spiracles and pectorals); belly a milky white. Its back-fin is placed nearer to the caudal than the hump on the Humpback, and in shape approaches a right angled triangle, but rounded on the forward edge, curved on the opposite one; the longest side joins the back in some examples, and in others the anterior edge is the longest. The gular folds spread on each side of the pectorals and extend half the length of the body.

"The habitual movements of the Finback in several points are peculiar. When it respires the vaporous breath passes quickly through its spiracles, and when a fresh supply is drawn into its breathing system, a sharp and somewhat musical sound may be heard at a considerable distance, which is quite distinguishable from that of other whales of the same family. We have observed the interval between the respirations of a large Finback to be about seven seconds. It frequently gambols about vessels at sea, in midocean as well as close in with the coast, darting under them, or shooting swiftly through the water on either side; at one moment on the surface, belching forth its quick, ringing spout, and the next moment submerging itself beneath the waves, as if enjoying a race with a ship dashing along under press of sail.

"In beginning the descent it assumes a variety of positions; sometimes rolling over on its side, at other times rounding or heaving its flukes out and assuming a nearly perpendicular attitude. Frequently it remains on the surface, making a regular course and several uniform 'blows.' Occasionally they congregate in schools of fifteen or twenty or less. In this situation we have observed them going quickly through the water, several spouting at the same instant. Their uncertain movements, however,—often showing themselves twice or thrice and then disappearing—and their swiftness, make them very difficult to capture. The result of several attempts to capture them was as follows; from the ship one was shot with the bomb gun, which did its work so effectually that although the boat was in readiness for instant lowering, before it got within darting distance the animal, in its dying contortions, ran foul of the ship, giving her a shock that was very sensibly felt by all on board, giving her a momentary heel of about two steaks. We had a good view of the underside of the whale as it made several successive rolls before disappearing, and our observations agreed with those made on the Sarah Warren in relation to the color and the creases on the throat and breast. The underside of the fins was white also. At another time the whale died about ten fathoms under water, and after carefully hauling it up in sight, the iron 'drawed' and away the dead animal went to the depth beneath. Frequently we have lowered for single ones that were playing about the ship, but by the time the boats were in the water nothing more would be seen of them.

"An instance occurred in Monterey Bay, in 1865, of five being captured under the following circumstances; a 'pod' of whales was seen in the offing, by the whalemen, from their shore station, who immediately embarked in their boats and gave chase. On coming up to them they were found to be Finbacks. One was harpooned, and, though it received a mortal wound, they all ran together as before. One of the gunners, being an expert, managed to shoot the whole five, and they were ultimately secured.

"Their food is of the same nature as that of the other rorquals, and the quantity of codfish that has been found in them is truly enormous." (Scammon).

Balænoptera acuto-rostrata davidsoni SCAMMON.
(For Prof. George Davidson.)
SHARP-HEADED FINNER WHALE.

Small; head pointed; pectoral fins small, pointed; baleen white; color blackish above, white below.

North Pacific Ocean.

"The name Sharp-headed Finner is applied to this, the smaller species of Balænoptera known on the coast. The only one we have examined was found dead on the northern shore of Admiralty Inlet, Washington Territory, by some Italian fishermen, in October, 1870, transported by them to the opposite shore and towed into Port Townsend Bay, where it was flensed on the beach. This opportunity of seeing the animal out of water was very interesting to us, for there was a mystery about its history that we had been unable to solve in twelve years observation, during which time we had traced it from the coast of Mexico to Behring Sea. In the strait of San Juan de Fuca opportunities were afforded for observing its havits more closely than elsewhere.

"The length of the individual captured in Admiralty inlet was twenty-seven feet. When compared with other *Balænidæ* it was so small that we were skeptical whether it was an adult or not, but, upon making an examination a well developed fœtus was found in it, five and one half feet long, which dispelled all doubts as to its maturity.

"The principal distinguishing features of this whale are its dwarfish size; its pointed head, which in form resembles a beak; its low, falcated dorsal fin, which is placed about two-thirds the length of the animal from the anterior extremity of its lower jaw, which is the longest; and its inordinately small, pointed pectorals, which are marked with a white band above and near their bases, and are placed about one-third the animal's length from its anterior extremity. The bone, or baleen, in its natural state is of a pure white, with a short, thin fringe of the same color. The number of laminæ on each side of the mouth was two hundred and seventy, and the longest of these measured ten inches. The surface of the animal was a dull black above, white below. The under side of both pectoral and caudal fins was white also.

"Seventy longitudinal folds extended along the throat and lower portion of the body, between and a little behind the fins, and while the outer surface of the folds was of a milky whiteness, the creases between them were of a pinkish cast, imparting the same shade to the throat as far back as the pectorals. The coating of yellowish fat that encased the body averaged three inches in thickness, and the yield of oil was about three hundred gallons.

"The habits of this whale are in many respects like those of the Finback. It frequently gambols about vessels while under way, darting from one side to the other beneath their bottoms. When coming to the surface it makes a quick, faint spout, such as would be made by a suckling of one of the larger cetaceans; which plainly accounts for whalemen taking it for the young of more bulky species." (Scammon).

Genus **Sibbaldius** GRAY. (For Robert Sibbald.)

Dorsal fin small, curved, pectoral fin small; head long; skin of throat with small longitudinal furrows; baleen short; cervical vertebræ free.

Sibbaldius sulfureous COPE. (Sulphur colored.)
SULPHUR- BOTTOMED WHALE.

Largest living mammal; brown or gray above; sulphur yellow below.

North Pacific Ocean.

"The largest whale found on the coast, and the largest known, is the Sulphur-bottom. Never having had an opportunity of obtaining an accurate measurement of its proportions, we can only state them approximately. Length sixty to one hundred feet. Its body is comparatively more slender than that of the California Gray. Its pectorals are proportionately small, even in comparison with the Sperm Whales which in size and shape they very nearly resemble, being short and rounded at their extremities. Its caudal fin bears about the same proportion to the body as does that of the Finback, while the dorsal is much smaller and nearer the posterior extremity. Its head is more elongated than that of the Finners. Its baleen is broader at the base; the color being a jet black in several specimens that we have examined, while others were of a bluish hue.

"Captain Roys, of whaling notoriety, has kindly furnished me with the following memoranda of a Sulphur-bottom Whale which was taken by him while he was in command of the barque Iceland. Length 95 feet; girth 39 feet; length of jawbone 21 feet; length of longest baleen 21 feet; yield of baleen 800 pounds; yield of oil 110 barrels; weight of whole animal, by calculation, 147 tons.

"The Sulphur-bottom, in its food and manner of feeding, is like the other whales of its kind. It is a true rorqual, with folds beneath the anterior part of the animal, which are a series of fine longitudinal furrows. The color of this, the greatest whale of the ocean, is sometimes lighter than the dull black of the lesser rorquals, in some instances it is a very light brown, approaching to white; but underneath it is of a yellowish cast or sulphur color, whence the name 'Sulphur-bottom' is supposed to have arisen. Its coating of blubber is unevenly distributed over its body, massively covering the top of the head, but more thinly covering the main portion of the trunk, while the posterior extremity, between the trunk and caudal fin is more heavily enfolded in the oily covering than all the rest.

"The Pacific species occurs at all seasons on the coasts of the Californias. During the months from May to September, inclusive, they are often found in large numbers close in with the shores, at times playing about ships at anchor in the open roadsteads, near islands, or capes, but in a general way they do not approach vessels with the same boldness that the Finback does, although we have observed them following a vessel's wake for several leagues. The Sulphur-bottom is considered the swiftest whale afloat. and for this reason is but seldom pursued, and still more rarely taken.

"On the second voyage of the Page six of these immense creatures were taken by the use of bomb-gun and lance off the port of San Quentin, Lower California, where the moderate depth of the water was favorable to the pursuit. Large numbers of them were found on this ground, where they were attracted by the swarms of sardines and prawns with which the waters were enlivened; and the whales, when in a state of lassitude from excessive feeding, would frequently remain motionless ten to twenty minutes at a time, thus giving the whaleman an excellent opportunity to shoot his bomb-lance into a vital part, causing almost instant death." (Scammon).

Dr. F. W. True, Curator, U. S. National Museum, thinks it is doubtful if this Whale is distinct from *Balænoptera physalus.*

Suborder **Denticete** (Tooth—Whale.)

Teeth nearly always present in the lower jaw and often in the upper; no baleen; rami of lower jaw united by a symphyseal suture; olfactory organ rudimentary or absent; nostrils combining in one spiracle.

Family **Physeteridæ.** (The Sperm Whale.)

Lower jaw slender and set with teeth, usually numerous, sometimes few; upper jaw large and toothless; head large; costal cartileges not ossified; skull usually unsymmetrical.

Subfamily **Physeterinæ.**

Lower jaw with numerous teeth, which are held in a long groove by a strong fibrous gum-like substance; upper part of cranium quite unsymmetrical through atrophism of the right nostril.

Two genera, each with one living species. They prefer warm waters and are found in all open seas except the Arctic and Antarctic Oceans. The food is squid, cuttlefish, octopus, etc. Sperm Whales are much less common now that formerly.

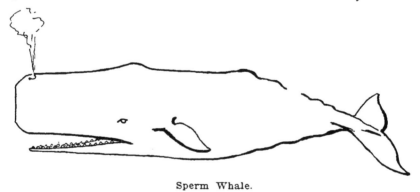

Sperm Whale.

Genus **Physeter** LINNEUS. (To blow.)

Dorsal fin obsolete; pectoral fin short, broad; head long and deep, squarish in front, with a large internal cavity filled with oil[1]; teeth 40 to 50; vertebræ 50; most of the cervical vertebræ united.

Physeter macrocephalus LINN. (Great—head.)
SPERM WHALE.

Blackish above; lighter below, particularly on the breast.

Length of adult male from 70 to 85 feet; females much smaller.

Found in nearly all seas from 56 degrees north latitude to 50 degrees south.

"This, the largest of the toothed cetaceans, is known to English and Amercan whalemen as the Sperm Whale, to the Germans as the Pottfish, and to the French as the Cachelot. It widely differs from all others of the order, both in figure and habits. The fully matured animal equals, if it does not exceed, the Bowhead in magnitude and in commercial value. The adult female, however, is only about one-third or one-fourth of the size of the largest male. She is likewise more slender in form.

"The largest males measure eighty to eighty-four feet. The pondrous head is nearly one third the whole bulk of the animal, and over one-quarter of its length. The opening of its mouth is about five-sixths the length of the head; the lower jaw, from the expansion of the condyles, contracts abruptly to a narrow symphasis, and is studded on each side with 22 to 24 strong, sharp and conical teeth, fitting into the furrow or cavity in the upper jaw, which is destitute of, or contains only rudimentary teeth. The tongue, which is usually of a whitish color, is not capable of much protrusion. The throat, however, is large, and is said to be capacious enough to receive the body of a man. The eyes are placed a little above and behind the angle of the mouth. A few inches behind the eyes are the openings of the ears, which are not one-fourth of an inch in diameter. Above, and at the junction of the head with the body proper, is a swell called 'the bunch of the neck.' About midway between this protuberance and the caudal fin is another and larger bunch, called the 'hump;' then follows a succession of smaller processes along the 'small' toward the posterior extremity which is called the 'ridge.'

"The pectorals or side fins are placed a little below and behind the eyes, and in size rarely exceed six feet in length and three feet in breadth. The caudal fin is about six feet in breadth, and measures from twelve to fifteen feet between the extremities, or about one-sixth of the length of the whole animal. Unlike the baleen whales the Cachelot has but one spiracle, or blow-hole, which is placed near the upper and anterior extremity of the head, a little on the left side; its external form is nearly like the letter S. This fissure in the adult is ten to twelve inches in length. The color of the Sperm Whale is black, or blackish brown above; a little lighter on the sides below, except on the breast, where it becomes a silvery gray. Some examples, however, are piebald.

"In the young Sperm Whales, as in the young of all cetaceans, the black-skin, or epidermis, is much heavier than in adults, it being a half an inch in thickness or thereabouts, while it does not exceed a quarter of an inch in the old whale. As age advances the skin becomes more furrowed. Beneath the black-skin lies the rich coating of fat or blubber, which yields the valuable oil of commerce. The head produces nearly one-third of all the oil obtained. Next to and above the bone of the upper jaw (which is termed the 'coach' or 'sleigh'), is a huge mass of cartileginous, elastic, tough fat, which is called the 'junk.' Above the junk, on the right side of the head, is a large cavity, or sack, termed the 'case,' which contains oil in its naturally fluid state together with the granulated substance known as 'spermaceti.' From this capacious hidden receptacle as much as fifteen barrels of head-matter has been taken. The 'ambergris' which is so highly prized, is nothing more than the retained anal concretion of a diseased whale. In the left side of the cranium, above the junk, is the breathing passage or nostril of the whale. This, with the case is protected by a thick, tough, elastic substance called the 'head skin' which is proof against the harpoon.

"We now come to the general habits of this gigantic animal, relative to its movements in the vast oceans of the globe. Among the whole order of the Cetaceans there is no other which respires with the same regularity as the Cachelot. When emerging to the surface, the first portion of the animal seen is the region of the hump, then it raises its head and respires slowly for about three seconds, sending forth diagonally a volume of whitish vapor like an escape of steam; this is called the 'spout,' which, in ordinary weather, may be seen from the masthead three to five miles. In respiring at leisure, the animal sometimes makes no headway through the water; at other times it moves slowly along at about the rate of two or three miles an hour; or if 'making a passage' from one feeding ground to another, it may accelerate its velocity. When in progressive motion (after 'blowing), hardly an instant is required for inspiration, when the animal dips its head a little and momentarily disappears, then it rises again to blow, as before, each respiration made with great regularity. The number of spoutings made when in a state of quietude depends on the size of the animal; varying in the adult female and in the young of both sexes from the largest and oldest males. The same may be said as to the length of time it remains upon or beneath the surface of the ocean. With the largest bulls the time occupied in performing one expiration and one inspiration is ten to twelve seconds, and the animal will generally blow from sixty to seventy five times at one 'rising', remaining upon the surface of the sea about twelve minutes. As soon as his 'spoutings are out' he pitches head foremost downward; then nearly perpendicular attitude, descends to a great depth, and there 'rounding out,' turns his flukes high in air, and, when gaining a remains from fifty minutes to an hour and a quarter.

"When a Cachelot becomes alarmed, or is sporting in the ocean, its actions are widely different. If frightened it has the faculty of instantly sinking, (as the sailors say, 'he can let go and go down in a jiffy'). When merely startled it will frequently assume nearly a perpendicular position, with the greater portion of its head above water, to look and listen; or when lying on the surface, it will sweep around from side to side with its flukes, to ascertain whether any object is within reach. At other times, when at play, it will elevate its flukes high in air, then strike them down with great force, which raises the water in spray and foam about it; this is termed 'lobtailing.' Oftentimes it descends a few fathoms beneath the waves, then, giving a powerful shoot nearly out of the water at an angle of forty-five degrees or less, falls on its side, or leaps bodily out in a semi-lateral attitude, coming down with a heavy splash, producing a pyramid of foam which may be seen from the masthead, on a clear day, at least ten miles, and is of great advantage to the whaler in searching for his prey. These singular antics of the Sperm Whale are said to be performed to rid itself of a troublesome parasite, known among sailors by the name of 'suckfish'; but the animal is seldom infested with the parasitic crustacea which are indigenous to the Rorquals and Right Whales.

"We may further add that it is one of the few species of the larger Cetaceans which inhabit every ocean not bound with icy fetters during the rigors of winter, and although great numbers of them are found in the cold latitudes; they also like to bask in the equatorial waters under a tropical sun. It is true, however, that but few are met with in the far northern limits of the Atlantic or the Pacific, compared with the numbers that inhabit the great range of the southern seas.

"The Cachelots are gregarious and they are often seen in schools numbering from fifteen to twenty up to hundreds. The oldest and largest males, however, for the greater part of the year roam alone; yet there is no lack of instances where these monsters have been found in herds by themselves; but the usual assemblage is made up of males and females—the latter with their young. At such times two or three large bulls are in attendance, which lead the van. The female is quite solicitous for her playful offspring, and when pursued the mother may be seen assisting it to escape by partly supporting it on one of her pectorals.

"The principal food of the Sperm Whale is familiarly named by whalers 'squid,' which includes one or more species of cuttle-fishes (cephalopods); occasionally the codfish, albicore, and bon-

ita are laid under contribution. But the true and natural way
in which this great rover of the hidden depths seeks and devours
its animal food is still tinged with mystery.

"The Sperm whale is usually found in the deep, open sea,
or, as whalemen term it 'off soundings,' but many instances are
known of their being seen in large numbers, and captures have
been made, on soundings. This has been the case to our knowl-
ledge off San Bartholome Bay and Ballena Bay, on the Lower
California coast, the depth of water varying in these places
from forty to eighty fathoms. Formerly this species was found
in great numbers along the coast of Upper California. The ships
cruising for them kept in a belt of water extending about one
hundred miles from the land and closing in with the shore."
(Scammon).

Family .Delphinidæ. (Dolphins, Porpoises, etc.)

Teeth usually numerous in both jaws; rostral portion of skull lengthened, about as long as cranium in some species, much longer in others; costal cartileges ossified.

This family includes the smaller cetaceans. Because of the lack of examples in museums the genera and species of this family are not well known and many changes in nomenclature may be expected as better material for study is acquired. There

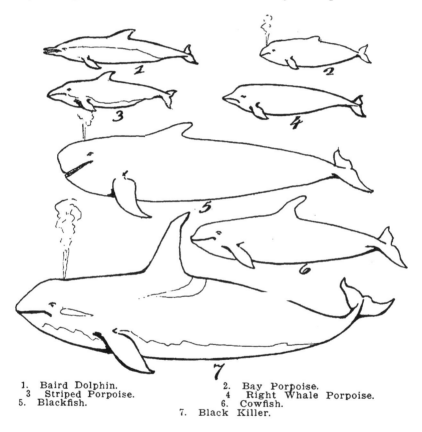

1. Baird Dolphin.   2. Bay Porpoise.
3 Striped Porpoise.   4 Right Whale Porpoise.
5. Blackfish.   6. Cowfish.
          7. Black Killer.

are fifteen or perhaps eighteen genera and probably more than fifty species. Many species are pelagic, others frequent the vicinity of shores, occasionally entering bays, while a few species are peculiar to large rivers, such as the Amazon and the Ganges. California gets a fair share of the marine species though several are almost exterminated.

These aquatic animals are active, voracious and usually gregarious. Their food is fish, squid, cuttlefish, etc. A few species are hunted for their oil, but many species yield too small amounts to make their pursuit profitable, particularly as their activity or peculiar habits make their pursuit difficult.

Subfamily **Delphininæ.**

Cervical vertebræ more or less consolidated; pterygoids not prolonged backward to articulate with the squamosals.

Genus **Lissodelphis** Gloger. (Smooth—dolphin.)

No dorsal fin; pectoral fin curved; depression in front of forehead moderate; rostrum long, tapering; teetr 43 to 47, small, sharp.

Lissodelphis borealis Peale. (Northern.)
NORTHERN RIGHT WHALE PORPOISE.

Form slender; beak short, distinct; flukes small; lower jaw longer than upper and curved upward at the extremity.

Length about 2200 mm. (87 inches); end of jaw to pectoral fin 625 (25); length of pectoral fin 300 (12); breadth of flukes 400 (16).

North Pacific Ocean. California. Japan.

"The Right Whale Porpoise of the western coast of North America, in habit and form, is nearly the same as the Right Whale Porpoise of the southern hemisphere (*peroni*), but it is not so beautifully marked in vivid contrast, in pure white and jet black, as the latter; the former being black above and lighter below, with but little of its lower extremities banded with white. The Right Whale Porpoise is not usually met with in large numbers, and is seldom found in shallow bays or lagoons. We have seen them as far south as San Diego Bay, on the Californian coast, and as far north as Behring Sea, showing plainly that the two species of the same genus have a feeding ground which em-

braces at least the western coast of North and South America."
(Scammon).

Genus **Phocæna** Cuvier. (Seal like.)

Dorsal fin rather small, varying in shape with species; pectoral fins ovate; rostrum short and broad; teeth 16 to 26, small, compressed; vertebræ 64 to 98.

Phocæna communis Lesson. (In common.)
BAY PORPOISE.

Head conical; body fusiform, slender; front margin of dorsal fin nearly straight, near margin concave; teeth 26—26; vertebræ 64 to 67; above slaty black; below lighter (male) or white (female).

Length about 1730 mm. (68 inches); end of jaw to dorsal fin 740 (29); height of dorsal fin 100 (4); length of pectoral fin 180 (7); breadth of flukes 320 (12.5).

North Atlantic and North Pacific Oceans, south to New Jersey and Mexico. Ascends rivers.

"This peculiar species of Dolphin is the least in size of the entire whale tribe inhabiting the Pacific North American coast. The body of the male is black above, a little lighter below; and while the female is of the same color above, it is lighter on the sides, with a narrow black streak running from the corner of the mouth to the pectoral, and the lower part of the animal is of a milky whiteness, yet the pectoral and caudal fins are black underneath or of a dark gray.

"The habits of this animal differ from those of other species found in the open sea or along the coast. Their favorite resort seems to be discolored waters between the limits of the pure ocean element and the fresh rivers. They are rarely seen far from either side of these boundaries. They are never found in large schools, but occasionally six or eight may be seen scattered about, appearing on the surface alternately, sometimes singly, or two or three together at the same insant. Neither do they make those playful heaps and gambols that the larger Dolphins do, their general habit being to make a quick puff and turn as soon as they appear above water, aparently choosing the darkness below rather than the light above. It is not from shyness, however, for they are met with about roadsteads and harbors, among shipping, and frequently play their odd turnings close to vessels under way, or at their moorings. By night, when at anchor, we have known them to play about the vessels rudder, but this may be regarded as an unusual occurrence.

"They feed on small fish, and are occasionally taken in the seines that are hauled along the shores of San Francisco Bay by Italian fishermen. The northern Indians frequently capture them about the inland waters during the calm, clear weather of the summer months." (Scammon).

Genus **Orcinus** Fitzinger. (A kind of whale.)

Dorsal fin very large and prominent; pectoral fins large, broad, ovate; head conical; rostrum broad, about as long as cranium; teeth 10 to 13, large; vertebræ 52; largest of subfamily.

Orcinus rectipinna Cope. (Straight—fin.)
STRAIGHT FINNED KILLER.

Dorsal fin at right angles to the body, extremely long, six feet or more in the male; no large white spot behind the eye; length of male about 20 feet.

North Pacific Ocean.

Orcinus ater Cope. (Black.)
BLACK KILLER.

Dorsal fin shorter, wider and curved backward; a white spot behind the eye.

North Pacific Ocean.

The following notes apply to both Killers, the description

and names being founded on these notes and Captain Scammon's drawings.

"The Orca—a cetaceous animal commonly known as the Killer—is one of the largest members of the Dolphin family. The length of the males may average twenty feet and the females fifteen feet. The body is covered with a coating of white fat, or blubber, yielding a pure, transparent oil. An extremely prominent dorsal fin, placed about two-fifths of the length of the body from the animal's beak, distinguishes it from all other Dolphins. In the largest species this prominent upper limb stands quite erect, reaches the height of six feet and frequently turns over sideways at the extremity. In the animals of more moderate size the fin is broader at the base, less in altitude, and is slightly curved backward, while upon others it is shorter still, and broader in proportion at its juncture with the back, and is more falcated. It is usually in color jet black above and lighter below; yet many of inferior size are most beautifully variegated, the colors being almost as vividly contrasted as the tiger of India. Some individuals have a clear white spot, of oblong shape, just behind the eyes, and a maroon band, of nearly crescent shape, adorning the back behind the dorsal fin, which it more than half encircles.

"The habits of the Killers exhibit a boldness and cunning peculiar to their carnivorous propensities. At times they are seen in schools, undulating over the waves—two, three, six or eight abreast—and with the long, pointed fins above their arched backs, together with their varied marks and colors, they present a pleasing and somewhat military aspect. Three or four of these voracious animals do not hesitate to grapple with the largest baleen whale; and it is surprising to see the leviathan of the deep so completely paralyzed by the presence of their natural, though diminutive enemies. Frequently the terrified animal—comparatively of enormous size and superior strength evinces no effort to escape, but lies in a helpless condition, or makes but little resistance to its merciless destroyers. The attack of these wolves of the ocean upon their gigantic prey may be likened, in some

respects, to a pack of hounds holding a stricken deer at bay. The
Orca, however, does not always live on such gigantic food; and
we incline to the belief that it is but rarely that these carnivora
of the sea attack the larger cetaceans, but chiefly prey with great
rapacity upon their young. The Orca finds its principal food in
the smaller species of its own family, together with the seals and
larger fishes. They will sometimes be seen peering above the
surface with a seal in their bristling jaws, shaking or crunch-
ing their victim, and swallowing them apparently with great
gusto; or, should no other game present itself, porpoises and
salmon may fill their empty maws, or a Humpback or Finback
Whale may furnish them with an ample repast.

"Compared with the other species of the Dolphin tribe the
Orcas are not numerous, neither do they usually go in large
schools or shoals, like the Porpoises and Blackfish. They are
seldom captured by civilized whalemen, as their varied and irregular movements make the pursuit difficult, and the product of
oil is even less than that of Blackfish, in proportion to their size.
By chance, however, we were so fortunate as to take one of
them, a female about 15 feet long, and on examining it to satisfy
ourselves about the character of its food, found that it consisted
of young seals. The covering of fat did not exceed three-fourths
of an inch in thickness, and was very white. The yield of oil
was one and a half barrels, and nearly as clear as springwater.

"The Killers I have noticed in the Gulf of Georgia, about
the northern end of Vancouver Island, and as far north as the
Aleutian Islands, appear to have more white on their sides and
are of a dull black on the back, the dorsal fin wider at the base
and shorter. I am fully convinced that there are two species at
least on the coast between the latitudes of 20 degrees and 60 degrees north; one with a dorsal fin excessively long, narrow at
base, standing very erect; the other species with a shorter dorsal
fin, somewhat curved, much broader and slanting backward."
(Scammon).

Genus **Globicephala** LESSON. (Ball—head.)

Dorsal fin long, low, curved; pectoral fins long, tapering curved; head globose; mouth oblique; rostrum very short and broad; teeth 7 to 11, large, in front half of jaws; vertebræ 57 to 60.

Globicephala scammoni COPE. (For Captain C. M. Scammon.)
SCAMMON BLACKFISH.

Size large; form stout; pectoral fins long, slightly curved, pointed; skull large and massive; color entirely black.

North Pacific Ocean.

"Blackfish are generally found wherever Sperm Whales resort, but in many instances they congregate in much' larger numbers, and range nearer the coast than the regular feeding ground of the latter. Although subsisting almost entirely on the same kind of food—the squid or octopus—still at times, when schools of them visit bays or lagoons, they prey upon the small fish swarming in those shallow waters. In Magdalena Bay we have seen them in moderate numbers, appearing as much at home as the Common Porpoise or the Cowfish. They collect in schools, from ten to twenty up to hundreds, and when going along on the surface of the sea there is less of the rising and falling movement than with the Porpoises, and their spoutings before 'going down' are irregular, both in number and time between respirations. If the animal is moving quickly much of the head and body is exposed. Whalemen call this going 'eye out.' In low latitudes in perfectly calm weather, it is not infrequent to find a herd of them lying quite still, huddled together promiscuously, making no spout and seemingly taking a rest.

"On the 14th day of December 1862, on the coast of Lower California, in latitude 31 degrees, land ten miles distant, a school of Blackfish was 'raised.' The boats were immediately lowered and gave chase, and three fish were taken. The largest was a male and measured as follows: Length 15 feet 6 inches. Depth

of body 3 feet 6 inches.   Circumference of body 8 feet 9 inches.
Expansion of flukes 3 feet 6 inches. Breadth of flukes 1 foot. From
end of head to spout hole 1 foot 6 inches. End of head to dorsal fin
4 feet 6 inches.   Length of pectorals 2 feet 10 inches.   Length
of spout hole across the head 4 inches.   The spout hole is of half
circular shape opening like a valve when the spout ascends, closing as it escapes.   The number of teeth on each side of the upper
jaw varies from ten to twelve, in the lower jaw from eight to
ten; the exposed parts from one-fourth to three-fourths of an
inch long.

"From all we can learn of their breeding habits they bring
forth their young at any time, or in any part of the ocean as
necessity may require.   Off the coast of Guatemala, in February,
1853, a calf taken from one was three feet long, the mother
measuring thirteen feet.   In the same school it was taken from
we saw several young ones about the same size as that above
mentioned.

"The Blackfish is taken for its oil, which is, however, much
inferior to that of the Sperm Whale.   The yield is small, from
ten gallons to ten barrels.   The blubber varies in thickness from
one inch to ten inches; its color is white.   The flesh of the Blackfish is like coarse beef, and after being exposed to the air a few
days, then properly cooked, is by no means unsavory food.   The
same may be said of the different species of Porpoises.   Formerly Blackfish were found in large numbers on the coast of Lower
California, but, probably from the same cause as made mention
concerning Sperm Whales, these grounds are now but little frequented by them."   (Scammon).

Genus **Grampus** GRAY.   (Great fish.)

Dorsal fin long, high, curved; pectoral fins long, narrow,
curved; head globose; mouth oblique; rostrum short and broad;
teeth two to seven, in front half of lower jaw only; vertebræ
68.

Grampus griseus Cuvier. (Gray.)
COMMON GRAMPUS.

Head globose with a suggestion of a beak; lower jaw shorter than upper; flukes narrow. *Adult*: back, dorsal fin and flukes dark gray or blackish, more or less tinged with purple; pectoral fins blackish mottled with gray; head and front part of body light gray tinged with yellow; belly grayish white; body marked with numerous irregular unsymmetrical streaks. *Young;* dark gray above; below grayish white; head whitish strongly tinged with yellow; sides with five or more narrow vertical stripes.

Length about 3200 mm. (125 inches); end of jaw to dorsal fin 1200 (48); height of dorsal fin 400 (16); breadth of flukes 720 (28); length of pectoral fin 600 (24).

North Atlantic and north Pacific Oceans. Mediterranean Sea. Japan. California.

Dall separated the Pacific Ocean animal under the name of *stearnsi;* but True cansiders them not seperable. The habits of Grampuses are similar to those of Porpoises.

Genus Lagenorhynchus Gray. (Flagon—snout.)

Dorsal and pectoral fins long and curved; rostrum large and broad; teeth 22 to 45; vertebræ 73 to 92.

Lagenorhynchus obliquidens Gill. (Oblique—teeth.)
STRIPED PORPOISE.

Form stout; beak very short; dorsal fin high, pointed, strongly curved; pectoral fins and flukes broad; teeth 31—31; vertebræ 74.

Length about 2200 mm. (87 inches); end of jaws to dorsal fin 915 (36); breadth of flukes 600 (24).

North Pacific Ocean. Puget Sound. California.

"This species of the smaller Dolphins varies but little in its general proportions from the Common Dolphin, except in its back fin which is more falcated and slender, and its snout, which is more blunt. In point of color it is greenish black on its upper surface, lightened on the sides with broad longitudinal stripes of white, gray and dull black, which in most examples run into each other, but below it is a pearly or snowy white. The posterior edge of the dorsal fin is tipped with dull white or gray, and sometimes the flanks are marked in the same manner.

"We have observed that this species has a wider range, congregates in larger numbers, and exhibits more activity than any other member of the dolphin family. They are seen, in numbers from a dozen up to many hundreds, tumbling over the surface of the sea, or making arching leaps, plunging again on the same curve, or darting high and falling sidewise upon the water with a spiteful splash, accompanied by a report that may be heard some distance. When a brisk breeze is blowing they frequentiy play about the bow of a ship going at her utmost speed, darting across the cutwater and shooting ahead, or circling around the vessel, apparently sporting at ease.

"The Striped Porpoises are often seen in considerable numbers about the large bays and lagoons along this coast, that have no fresh water running into them. They abound more along the coasts where small fish are found than in midocean, as they principally prey upon the smaller finny tribes; and to obtain them shoot swiftly through the water, seizing the object of their pursuit with the slightest effort. Occasionally a large number of them will get into a school of fish, frightening them so that they will dart around in all directions, and finally get so bewildered as to loose nearly all control over their movements. At such times the Striped Porpoise is manifestly the 'sea swine', filling itself to repletion." (Scammon).

Genus **Delphinus** Linnæus. (Dolphin.)

Dorsal and pectoral fins long, rather narrow, curved; a distinct depression across the head in front of forehead; rostrum nearly twice as long as cranium, narrow; teeth 47 to 65, narrow, small; vertebræ 73 to 76.

Delphinus delphis LINN. (Dolphin.)
COMMON DOLPHIN.

Form and disposition of color markings very variable; teeth
47—46 to 50—51; length about 2265 mm.   (90 inches) ; end of
jaw to dorsal fin 1000 (40) ; end of jaw to pectoral fin 500 (20) ;
height of dorsal fin 230 (9) ; length of pectoral fin 350 (14) ;
breadth of flukes 520 (21).

"Mr. Dall was unfortunately unable to compare his skele-
ton with that of *D. delphis,* to which species *D. bairdii,* if distinct,
is undoubtedly most closely allied. From the evidence now obtainable I am unable to distinguish between *D. delphis* and *D.
bairdi,* and must therefore regard the latter as identical with
the former." (True).

Pelagic. Found in most seas. I have a skull picked up on
the beach near San Diego, and have seen others. The following
is from Captain Scammon's account of *Delphinus "bairdii."*

"This Dolphin inhabits the Pacific North American coast,
in common with other varieties which abound in these waters.
At a distance it much resembles the Common Porpoise of fishermen and sailors; but it differs in several points from that species.
We were so fortunate as to obtain two female specimens off
Point Arguello, in the fall of 1872, from which we obtained the
following notes. Apparently both individuals were adults, and
nearly the same size and weight. The body of *Delphinus bairdii*
is more slender, and its snout more elongated and rounded, than
that of the Striped or Common Porpoises, and may be compared
to the bill of a snipe. Its teeth are slender, conical and slightly
curved inward. Its dorsal fin is more erect and less falcated
than that of the *Lagenorhynchus obliquidens,* while its pectorals
are nearly the same shape and comparative proportions; but the
caudal fin is less in breadth and greater in proportionate expansion. Its back, immediately forward of the dorsal fin, is somewhat concave, so that when taking a side view the upper contour
appears lower before than behind the fin. Its varied colors are,
top and sides of head black, sides of body behind the vent and

both sides of pectorals and flukes a greenish black; a black patch around the eye with a black streak passing forward above the mouth, a continuous black streak from the side of the under jaw to the anterior edge of the pectorals. Sides behind the eye gray, the upper boudary of this color being somewhat above the plane of that organ, beginning to curve downward just behind the dorsal fin, and meeting both black and white marks between the vent and flukes, in or near the mesial line of the under side of the body. A lanceolate white patch extending on the ventral side from the middle of the jaw to the vent. A narrow white stripe extending from the corner of the mouth backward, on each side, slightly arched above the pectoral and then curving downward gradually, the two meeting below in the region of the vent. Another, still narrower and somewhat obscure, starts at the same place as the last, but is soon lost in the white ventral patch before alluded to. The *Delphinus bairdii* may be considered symmetrical in its proportions. It moves through the water with great swiftness and grace. Appended are the dimensions in feet and inches, of the examples above mentioned.

Total length of the animals ....6'—7" 6'—9"
Anterior edge of pectorals ....1'   1'
Expansion of flukes ....1'—6"   1'—5"
Height of dorsal fin ....   7"   7"
Circumference before the dorsal fin ......3'—4"   3'—3"

Genus **Tursiops** Gervais.   (Dolphin like.)

Dorsal and pectoral fins long, narrow, curved; sides not banded; a distinct depression across head in front of forehead; rostrum moderately long, tapering; teeth 22 to 26, large, smooth; vertebræ 61 to 64.

Tursiops gilli Dall.   (For Prof. T. N. Gill.)
COWFISH.

"Exterior known only from an outline drawing and record

of two momentary observations by Scammon. Teeth 22—22. Habitat; North Pacific Ocean, Monterey, California and Lower California." (True).

"This Porpoise is larger than the Striped or Right Whale species, and is known by the name of Cowfish. It is longer also in proportion to its girth, and its snout is somewhat contracted. Its teeth are much larger, straight, conical, and sharply pointed, but less in number. A specimen taken at Monterey in 1871, had 24—23, 24—23. The animal also differs in color, it being black all over, lightened a little below.

"The habits of the Cowfish, as observed on the coast of California and Mexico, are strikingly different from those of the true Porpoise. It is often remarked by whalemen that they are 'a mongrel breed' of doubtful character, being frequently seen in company with Blackfish, sometimes with Porpoises, and occasionally with Humpbacks when the latter are found in large numbers on an abundant feeding ground. They are met with likewise in the lagoons along the coast, singly, in pairs, or fives, or sixes, rarely a larger number together, straggling about in a vagrant manner in the winding estuaries, subsisting on the fish which abound in these circumscribed waters. At times they are seen moving lazily along under the shade of the mangroves that in many places fringe the shores; at other times lying about in listless attitudes among the plentiful supplies of food surrounding them." (Scammon).

Order **Ungulata**.

Toes more or less completely enclosed by horny hoofs or with broad claws; no clavicles; molar teeth with ridged or tuberculated grinding surfaces.

Suborder **Artiodactyla**.

Feet cleft; first toe wanting; second and fifth toes small, rudimentary or absent.

Superfamily **Pecora**.   Ruminants.

Stomach with four compartments; food regurgitated and remasticated; horns or antlers usually present; upper canine teeth usually absent, sometimes present and occasionally largely developed.

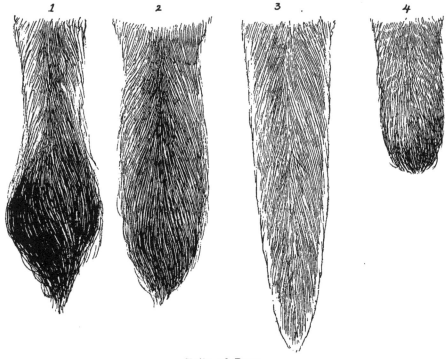

Tails of Deer.

1. Mule Deer.   2. Black-tailed Deer.   3. Virginia Deer   4. Wapiti.

Family **Cervidæ.** (Deer.)

Male usually, female rarely, with deciduous bony antlers placed on a permanent short pedestal; upper canines usually present, sometimes highly developed in males of certain Asiatic species; upper incisors absent; second and fifth toes present but small; no gall bladder.

Genus **Cervus** LINNEUS. (Deer.)

Antlers, on male only, two or three times as long as head, usually round, branched, the tines turned forward, brow tine low; posterior nares not divided; canines never projecting beyond edge of lips; lachrymal pit large; ears rather small; no interdigital "glands"; hoof rounded, oxlike in form; metatarsal gland present on hind leg; tarsal gland absent; tail short.

Dental formula, I, o—4; C, 1—o; P, 3—3; M, 3—3×2=34.

Cervus roosevelti MERRIAM. (For Theodore Roosevelt.)
ROOSEVELT WAPITI.

Male; size large; skull and antlers massive; beams of antlers relatively short and straight, with terminal prong aborted; most of face black or brownish black; hairs of neck long and forming a mane on the throat; a dusky or black stripe on top of neck; extending a greater or less distance on the back, remainder of neck brown; breast and belly dull chestnut brown; sides and back grayish brown; a large pale tawny patch on the rump. *Female;* no antlers; smaller; dark colored parts paler.

Length of adult male about 2500 mm. (98 inches).

Type locality, Olympic Mountains, Washington.

Pacific coast from northwestern California to British Columbia. When the first white men came to California Wapiti of this or the next species were common in many places in the central and northern parts of the State. Now this species is limited to a few inaccessible places in the three or four northwestern counties.

"Wapiti" appears to have been the Iroquois name of the animal commonly called the American "Elk." The European Elk is closely related to the American Moose, while the European analogue of the Wapiti Deer is the Red Deer or Stag; hence "Elk" is misapplied as a name for the American animal, and Wapiti, as the next best known name, should be used.

Wapiti prefer forests moderately free of undergrowth, in mountainous or hilly regions. The food is coarse and varied, consisting largely of leaves and twigs. They are good trotters and usually adopt that gait for rapid traveling unless very much hurried, when they break into a fast run. This gait an old fat buck cannot sustain long before coming to bay, but poor or young animals can run a considerable distance. The voice is high, sharp and forcible, but is only used in defiance or in great alarm.

Wapiti are somewhat gregarious and are occasionally seen in large herds in the Rocky Mountains. They are polygamous, the strongest bucks gathering a small band of does in the rutting season and driving away weaker rivals. The rutting season of the eastern species is September and the fawns are dropped about May; probably the same dates hold good for our species. The bucks are tyrannical to the members of their harem. Twins are infrequent. The venison of Wapiti is not as tender as that of the smaller Deer, but it is very nutritious. It is very difficult to preserve. Still hunting on foot is the usual method of hunting Wapiti, and in northwestern California this is practically the only method available.

Cervus nannodes MERRIAM. (Small.)
CALIFORNIA WAPITI.

Size small; legs short; coloration pale; head, neck and shoulders grizzled grayish brown; back and flanks varying from buffy gray to grizzled buffy whitish; front of legs and feet light tawny; rump patch white, small and narrow.

The type, a two year old male, measured, length 2030 mm. (80 inches) ; tail vertebræ 140 (5.50) ; hind foot 620 (24.40).

Type locality, Buttonwillow Ranch, Kern County, California.

The California Wapiti are now limited to a small band running in Kern County. This is the pitiful remnant of the thousands that ranged over the San Joaquin and Sacramento Valleys when the first gold hunters came to this region. The chief of the U. S. Biological Survey (Dr. Merriam) made an unsuccessful attempt to place this band in the Sequoia National Park in the autumn of 1904. It is to be hoped that some means will be found to preserve the few individuals left. This seems to be a small, valley-loving species, and is not known to have occurred outside of California. So far as known, their habits, are like those of other Wapiti, except that they often frequent marshy localities.

Genus **Odocoileus** RAFINESQUE. (Tooth—hollow.)

Antlers, on male only, less than twice as long as head, round, branched, not palmated, brow tine some distance above base of antler; posterior nares divided by a bony septum (vomer) ; upper canines absent; lachrymal pit large; ears medium or large; interdigital "glands" present; metatarsal and tarsal glands present on hind leg; hoof narrow and pointed; tail of medium length.

Dental formula, I, 0—4; C, 0—0; P, 3—3; M, 3—3×2=32.

Odocoileus hemionus RAFINESQUE. (Mule.)
MULE DEER.

Antlers usually dividing in two subequal forks and each fork disposed to branch again; tail vertebræ shorter than the ear; a strip of naked skin on underside of tail; metatarsal gland (on outside of hind leg) five to six inches long; ears very large. *Winter pelage;* dark gray above, fading as the season advances; breast blackish; a large patch surrounding the tail, from the rump to between the legs, dull white. *Summer pelage;* yellowish brown to reddish brown. *Young;* brownish yellow more or less regularly spotted with dull white.

Length of male about 1575 mm. (62 inches); tail vertebræ 185 (7.25); hind foot 475 (18.65); ear from crown 240 (9.50). Length of female about 1450 (57); tail vertebræ 175 (6.90); hind foot 445 (17.50); ear 225 (8.85).

(Note: the length of the hind foot is the distance from the point of the longest toe to the extremity of what is popularly called the "knee" which is really the true heel. Ungulates walk on the ends of their toes).

Type locality, upper Missouri River.

The Mule Deer ranges over a large part of the United States, from northern Arizona to British America, and from the great plains to the Sierra Nevada and Cascade Mountains. In the southern part of this range the true Mule Deer blends with the two succeeding subspecies. It is moderately common on the eastern slope of the northern part of the Sierra Nevada. It prefers the foothills of mountain ranges and broken ground in plains, but is also found in mountains. The gait of the Mule Deer is less graceful than that of the Virginia Deer. The run is a series of high bounds, rapid but too tiresome to be sustained long.

The Mule Deer is easily distinguished from the Black-tailed Deer and Virginia Deer and its western forms by the much smaller tail, which is naked part way down on the under side and has the terminal third black and the remainder white. The white hairs wear away easily and frequently the middle of the tail is very slender. The metatarsal gland is the longest found on any North American deer. The bare strip is easily seen by parting the hairs over it, these hairs being longer than those of the remainder of that side of the leg.

The antlers are different from those of the white-tailed group of deer in one respect; those of the latter species have an indeterminate number of tines, aged bucks having numerous tines, though these are only in a general way an index of his

age; while the antlers of the Mule Deer seldom have more than
ten points, including the brow tines. Eight to twelve inches
from the base each antler forks, and about six inches further
each branch usually forks again in middle aged bucks. This
is the normal adult form of antlers of the Black-tailed Deer and
all the subspecies of the Mule Deer. Now and then a buck adds
a tine or two, but these are not common. The antlers of deer
are not composed of horn but are bone, and it is a mistake to
speak of antlers as "horns." They are grown underneath a skin,
much as other bones are; not from their bases and inner surfaces
as horns are.

The Mule Deer, as well as our other species, eats a variety
of plants, prefering a considerable proportion of twigs and foliage
of shrubs and trees intermixed with grass and other plants, as
well as seeds, fruits and such nuts as they can chew, such as
acorns. In localities where they are not distrubed they feed
more or less in the daytime, but where they are hunted they become principally nocturnal.

Odocoileus hemionus eremicus MEARNS. (Hermic, i. e., a dweller in the desert.)
BURRO DEER.

Similar to *hemionus;* larger; paler; in winter yellowish
drab gray, darkest on the back, palest on the sides; breast sooty
drab; sometimes the dark area from the rump extends a short
distance down on the tail, but more often it is as indicated in the
drawing, which was made from a recently killed buck, near Black
Mountain on the Colorado Desert, December 10th. This was a
medium sized buck and measurements were as follows: total
length 1680 mm. (66 inches); tail vertebræ 190 (7.50); hind
foot 491 (19.30); ear from crown 250 (9.80); girth of body
behind fore leg 1050 (41.30). I estimated his weight at 150
pounds but others of the party thought he was heavier. A female
killed a few days previously in the same locality measured, total

length 1430 (56.30); tail vertebræ 180 (7.10); hind foot 430 (17); ear from crown 218 (8.60); girth 390 (35).

Type locality, northwestern Sonora, Mexico.

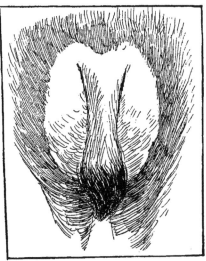

Tail of Burro Deer.

Burro Deer are seldom seen in the mountains, but are found along their base, and in comparatively level land, even in the mesquit timber of the Colorado Valley, where they feed on willow twigs along the sloughs. I saw also where these Deer had eaten the wild gourds ("mock oranges"). They are not found many miles from water, as in warm weather they visit ponds or streams nightly to drink, and in cooler weather every second or third night. They are found in small parties, sometimes singly, a dozen being a large band as far as my observations go. The antlers are commonly very regularly double-forked.

Odocoileus hemionus californicus CATON.
CALIFORNIA MULE DEER.

Similar to *hemionus;* considerably smaller; color more tawny; tail rather longer proportionally and usually with a distinct brownish or blackish stripe on the upper side from the rump to the black terminal switch, which often has a light brown or whitish stripe underneath. The following measurements were taken from an average pair. This buck weighed about eighty pounds. He was shot December 23rd, and had recently dropped his antlers, the scar not being fully healed. The Indian who shot him told me that another buck of the same band was still carrying his antlers. Length of male 1410 mm. (55.50 inches); tail vertebræ 172 (6.77); hind foot 440 (17.30); ear from crown 230 (9); height at shoulder 890 (35); at hip 1005 (40). Length of female 1340 (52.70); tail vertebræ 156 (6.15); hind foot 398 (15.67); ear from crown 312 (8.38).

Type locality Gaviota Pass, Santa Barbara County, California.

The range of the California Mule Deer appears to be limited to southern California and northwestern Lower California. Probably intergradation occurs with true *hemionus* in the southern Sierra Nevada. Their range touches that of the Burro Deer along the edge of the Colorado and Cocopah Deserts, where the great difference in size of the two subspecies becomes very apparent. They are found in the mountains and foothills and to some extent in the brushy valleys. They frequent pine and oak timber and the chemisal of the hillsides. Their antlers are less regular in form than those of the Burro Deer. The new antlers begin their growth sometimes as early as the middle of May, in other cases as late as the end of June. In about two months the growth is finished and soon after the buck begins to strip off the skin ("velvet") by rubbing the antlers against trees and bushes. The antlers are worn until the rutting season is over, and are dropped from November to January, the time depending on the age of the buck, locality and other circumstances. If a buck is castrated, accidentally or intentionally, the growth of the next set of antlers is more or less imperfect, the skin is retained and the antlers do not drop off as usual, but become permanent. The next season abnormal points are grown, mostly about the bases of the antlers and in the course of years the antlers become a mass of points, mostly small and still covered with harsh skin. In cold climates such antlers freeze and the points are often broken off. I have seen a set of these abnormal antlers from a California Mule Deer killed in Riverside County, in which none of the points appeared to have been broken off, but none were of greater length than the ears.

The fawns appear to be dropped principally in July. , Twins are frequent. The hearing ˌ keen. The sight is comparatively poor. The scent is delicate and acute and is depended on nearly as much as the hearing for warning.

Odocoileus columbianus RICHARDSON.   (Of the Columbia River.)
BLACK-TAILED DEER.

Size similar to that of the White-tailed Deer and scarcely larger than the California Mule Deer; antlers similar to those of the Mule Deer; tail longer than that of the Mule Deer and considerably broader, covered with hair underneath, the under side white, the upper side brown on the basal half, and dull black on the remainder; naked strip of metatarsal gland (on outside of hind leg), two or three inches long; ear smaller than that of the Mule Deer; body and legs short. *Winter pelage;* above gray, more or less tawny, darker on the upper side of the neck; a dark streak on the under side of the neck, becoming black on the breast, shading to brown on the belly; between the thighs is a white area extending to the tail. *Summer pelage;* above yellowish red or dull reddish·brown.

Type locality, the lower Columbia River.

The "Black-tailed Fallow Deer" as they called it was first described by Lewis and Clark in the report on their memorable Expedition. The range of the Black-tailed Deer is from the northern Sierra Nevada and Cascade Mountains to the Pacific Ocean, north to British Columbia.

Townsend says that the Black-tailed Deer in northern California migrate, leaving the foothills in spring and going high in the mountains in summer, even to timber line on Mount Shasta; returning in autumn.   Belding noticed a similar migration of Deer in the central Sierra Nevada.

Townsend says that the new growth of antlers begins early in April and is completed in July.   The antlers drop in January. The rutting season is about November, and the fawns are dropped in May and June.   Twins are the rule.   The fawns are bright bay spotted more or less regularly with white.   In that part of California north of San Francisco and west of the Sierra Nevada only Black-tailed Deer occur.

Odocoileus columbianus scaphiotus Merriam. (Boat— ear.)

SOUTHERN BLACK-TAILED DEER.

Similar to *columbianus* but ears larger and much broader; colors paler; teeth larger and heavier.

Length of type specimen (a male) 1465 mm.   (58 inches); tail vertebræ 135 (5.30); hind foot 452   (17.75); ear 178 (7); breadth of ear 106 (4.15).

Type locality, Laguna Ranch, Gavilan Mountains, San Benito County, California.

The range of this   subspecies is the coast region south   of San Francisco Bay so far as is known at present. The southern limit is not known.

PRONG-HORN ANTELOPE

Family **Antilocapridæ.** (American Antelopes.)

Horns deciduous, hollow, recurved. with a flattened prong in front; horn cores bony, not branched, flattened; orbit close beneath base of horn; no lachrymal pit; no tarsal or metatarsal glands; secònd and fifth toes absent; interdigital glands present; cutaneous glands present under· each ear, on the rump, on each hip, and behind each hock; gall bladder present; mammæ four; hairs long, hollow, coarse and brittle; pelage not differing with age, sex or season to any material extent.

This is one of the smallest families of mammals, consisting of but one genus with a single species, though this will probablv bear subspecific division. The distribution is North American. Gramnivorous, digitigrade, terrestrial and principally diurnal.

Dental formula, I, o—4; C, o—o; P, 3—3; M, 3—3×2=32.

Genus **Antilocapra** Ord. (Antelope—Goat.)

Body short; ears of moderate length; eyes .very large; hairs on top of neck long, forming a mane.

Antilocapra americana Ord.
PRONG-HORN ANTELOPE.

Horns fully developed in male only, those of the female rudimentary, not much longer than the surrounding hairs; narrow transverse band between the eyes, top and sides of muzzle and a patch beneath each ear (wanting in the female) brownish or blackish; edges of upper lip, chin, sides of face, spot behind the ear, a narrow crescent on the upper part of the neck, a triangular patch below this, a large square patch on the rump including the tail, belly and lower half of the sides white; remainder of upper parts and legs russet yellow or yellowish brown; hoofs, and horns except tips, black.

Length of male about 1500 mm. (59 inches); tail vertebræ, 125 (5); height at shoulders 840 (33); at hips 940 (37).

Prong-horn Antelopes formerly ranged over most of the

untimbered parts of the United States west of the Mississippi River, northern Mexico and the southern part of British America. Prior to the discovery of gold in California they were abundant in the San Joaquin-Sacramento Valley and other parts of the State. In 1877 I saw a band of about two dozen where Perris, Riverside County now stands, and the next year I saw one within the limits of what is now the city of Riverside. At this writing they are almost exterminated in this State. There are a very few in Modoc, Lassen and Mono Counties, and a small band or two in the deserts in the southeastern part of the State. All told there may be two or three hundred left and this number is steadily diminishing.

Prong-horned Antelopes are found in open treeless regions, very seldom among trees, never in dense forests. They often frequent broken and hilly ground. Their food is mostly grasses, seldom twigs or leaves of bushes and trees, in this respect being unlike the deer family. Their run is very rapid, probably faster than that of any deer and they can continue this rapid run longer than the deer. They are able to jump across wide ditches, making very long horizontal leaps, but are unable, or do not know how, to leap over obstacles three or four feet high.

The rutting season is September and the young are dropped in May. Twins are frequent. Prong-horns are easily tamed but are difficult to keep in good health. Their power of scent is acute, but their sight and hearing is only moderately good. They are shy and timid, but are inquisitive, their curiosity being taken advantage of by hunters to entice them within gunshot range. Their flesh is not very palatable when freshly killed but becomes better with a day or two's keeping. It is less nutritious than venison. Prong-horns are more gregarious than deer at most seasons. They are called Cabree by the French Canadians.

Family **Bovidæ**. (Cattle, Sheep, Antelopes and Goats.)

Horns usually present, permanent, hollow and placed on a bony core; canines absent; second and fifth toes present in some genera, absent in others; gall bladder usually present.

There are thirty or thirty-five genera of Bovidæ and about one hundred and fifty species. Bovidæ are digitigrade. gregarious and principally diurnal. They are principally Old World in distribution. Several species are domesticated and have been introduced in all civilized countries. Their food is vegetable, mostly herbs and their seeds. As is the habit with most Ungulata, the food is gathered in with the tongue; pressed by the lower incisors against the pad-like end of the upper jaw and torn loose with a pull. It is then swallowed with but little chewing; and later remasticated. The Bovidæ form a large source of food supply for the human race, and a considerable part of their clothing also.

Genus **Ovis** Linneus. (Sheep.)

Horns present in the males, usually large, curved backward spirally; females usually with small horns; a small lachrymal pit usually present; interdigital glands in all the feet.

Dental formula, I, 0—4; C, 0—0; P, 3—3; M, 3—3×2=32.

The domestic sheep has been introduced into nearly all parts of the world by man. Its origin is unknown. It is unlike any of the now known wild sheep, none of which have such heavy coats of wool. There are eighteen species of wild sheep now known, the greater number being Asiatic.

Ovis canadensis Shaw. (Of Canada.)
ROCKY MOUNTAIN BIGHORN—MOUNTAIN SHEEP.

General color grayish brown; nose and chin lighter; belly and a large patch on the rump and about the tail white, tail and a narrow stripe on the rump like the back, horns of male massive.

Type locality, Rocky Mountains of Alberta, Canada.

Rocky Mountains to Sierra Nevada and Cascade Mountains and intervening ranges. In California Rocky Mountain Bighorns were formerly found in parts of the Sierra Nevada and on Mount Shasta, but they are apparently now exterminated in those mountains. It is possible that these animals were not *canadensis,* but were *nelsoni* or some unnamed form. Material is lacking now to determine this point, with little probability of more being obtained. Two or three very small bands still exist in certain mountains of southwestern California that are probably intermediate between the above species and *nelsoni.* Poachers are destroying them and their destruction is probable in a few years.

Ovis nelsoni MERRIAM. (For E. W. Nelson.)
NELSON BIGHORN.

General color above varying with season and locality from pale ashy gray or pale dingy brown to dirty white; rump patch and back part of hams white; belly white; breast sometimes brownish white but often slate gray; fore part of legs brownish gray; tail and a narrow stripe on the rump drab gray and often a drab gray stripe from the neck over the withers.

Length of male about 1525 mm. (60 inches); tail vertebræ 125 (5); hind foot 400 (15.75); ear from crown 130 (5.15); length of horn around curve 700 to 900 (27 to 35). Length of female about 1400 (55); tail vertebræ 110 (4.33); hind foot 375 (14.75); ear from crown 130 (5.15); horn around curve 280 (11).

Type locality, Grapevine Mountains near Death Valley, California.

The range of the Nelson Bighorn appears to be southern Nevada, southeastern California, the northeastern border of Lower California and probably western Arizona. An adult male Nelson Bighorn in good condition will weigh two hundred and fifty pounds and a female one hundred and fifty. They prefer hilly or mountainous regions, preferably arid with occasional springs or waterholes. I have seen tracks in small valleys but they do not often come down on level ground. If not disturbed they do not frequent exceptionally rough mountains, but when hunted much they get into the roughest places they can find.

Their food is principally "browse", i. e., leaves and twigs of shrubs. In some of the desert mountains a very coarse perennial grass known by the Mexican name of "galletta" grows, which they eat. The contents of a stomach of a Bighorn which I killed in the Providence Mountains in June consisted principally of leaves, twigs and flowers of *Rhamnus, Ephedra* and *Rhus,* with some unripe fruits of the *Rhus,* and a little grass. Bunch grass was green and plentiful at the time there but evidently the Bighorn preferred the coarser food. The Indians tell me that the Bighorns eat the larger species of cacti when water is scarce.

Bighorns vary in their drinking habits with locality and season. In the desert mountains in summer they drink daily if practicable, coming to water most often about the middle of the afternoon, but sometimes in the forenoon. In cool weather they drink less frequently, and even in summer those running in cool mountains do not drink often. A small band running in the Providence Mountains at 5,000 to 7,000 feet altitude in 1902 did not appear to go to water more than two or three times a month. The spring was down in the hot foothills and inconvenient, while on the crest of the mountains the weather was cool and feed abundant and green.

Bighorns seem to be crepuscular and diurnal in habit, but if disturbed often they feed some in the night. They appear to lie down in the forenoon, sometimes soon after sunrise. In warm localities the beds are pawed out in the shade of shrubs or rocks, but in cool mountains they are made in open places commanding a clear view around them.

The voice is said to be similar to that of the domestic sheep, but coarser. It is probably used but little. A Bighorn ewe lamb, three or four weeks old, that I had alive a few days, bleated much like a domestic lamb, but coarser and nôt as loud, nor as frequently. I had hopes of raising this lamb but did not succeed in getting into the settlements in time to save her. She was very gentle and soon accepted our company in lieu of her own kind. If we all went out of sight of where she was tied she soon got uneasy and bleated, but when we came back she settled down contentedly.

Bighorns are exceedingly sure footed animals and quite active. They do not seem to run fast, and I doubt if they could run far at their most rapid gait. The soft, rubber-like soles of their hoofs do not slip on smooth rocks. In jumping upward they can surpass any deer, and they will go rapidly down a cliff where it would seem impossible for anything not provided with wings to pass. The old stories of Bighorns jumping over cliffs and alighting on their horns are untrue. In jumping downward they alight on their feet, and the ewes are as active and sure footed as the rams. The horns of old rams are more or less bruised and worn away at the points by striking against rocks in feeding and in passing along cliff sides.

My impression is that Bighorns are more easily killed than deer. i. e., a wound that a deer would probably recover from would probably prove fatal to a Bighorn. I consider the mutton of Bighorns equal to the best venison in flavor, but the few Bighorns whose flesh I have had the opportunity to taste were all in good condition.

The lambing season is principally March. I have never seen twins and do not know of any record of more than one lamb at a time. The principal natural enemies of Bighorns are pumas and coyotes. Indians kill many, but the white hunters are responsible for the extermination of Bighorns over much of their former range. They seem to be able to hold their ground better in the comparatively open hills of the deserts than in the high timbered mountains.

NELSON BIGHORN

Genus **Oreamnos** Rafinesque. (Mountain—lamb.)

Horns, present in both sexes and nearly of the same size, black, slightly curved backward; spinous processes of interscapular vertebræ very long and rigid; hair very long, under fur short, wooly; a beard-like tuft under the chin.

Oreamnos montanus Ord. (Of the mountains.)
MOUNTAIN GOAT.

Hoofs, horns and edges of the nostrils black; pelage everywhere dirty white; smaller in size than the Bighorn; horns 150 to 200 mm. long (6 to 8 inches).

Type locality, Cascade Mountains near the Columbia River.

Higher peaks of the Cascade Mountains. Said to have been found in the northern Sierra Nevada but not now known to occur there. Newberry says, in speaking of Sheep Rock, Mount Shasta: "It is said that the Rocky Mountain Goat is also to be found there, but of that I have very great doubt." Captain Charles Bendire recorded it from Inyo County in 1868. It is practically certain that the species is now not living in this State.

This animal has the habits and somewhat the form of a goat; nevertheless it is an antelope and a near relative of the Swiss Chamois. A better name for the animal would have been American Chamois. They frequent the higher parts of rough mountains, and are said by some authors to be very watchful and difficult to hunt, while others say it is stupid and easily shot when the hunter succeeds in climbing to the rugged peaks which they frequent. In winter they descend to more moderate elevations to obtain food. The young are dropped in June.

Order **Glires**. (The Rodents or Gnawers.)

Incisors two in lower jaw, usually two but occasionally four in upper. large, with chisel-shaped points, fitted for gnawing; no canines, but a considerable gap in their usual place; premolars present in some families, absent in others; molars usually three in each side of each jaw, adapted for grinding; condyles of lower jaw not received in special sockets, but permitting more or less longitudinal grinding movement of the jaw; cerebrum small, but little convoluted; clavicles present but sometimes rudimentary; digits generally five, furnished with nails or claws; food chiefly vegetable; modes of life greatly diversified.

Rodents form the largest order of mammals, containing nearly or quite one thousand living species.   It is also the most widely distributed terrestrial order.   South America seems to be the center of distribution.

Suborder **Simplicidentata**.

But one pair of upper incisors; enamel coating incisors confined to their front surfaces; incisive foramina distinct and of moderate size.

The general structure of the various genera of this suborder are so similar that the characters available for distinguishing them are comparatively trivial and of slight structural importance.

Family **Sciuridæ**. (Squirrels.)

Skull varying with genera in length relative to breadth; postorbital processes present, various in form; first premolar small, often deciduous; molars rooted, tubercular; palate broad; clavicles developed; fibula free; tail without scales, well haired, various in length of vertebræ and hairs; ears varying in length from quite long to rudimentary.

The Squirrels are a large and important family of rather small sized mammals.   They are distributed over nearly all parts of the world except Australia, and are well represented in California. The larger species are hunted for their flesh. Many species are destructive of crops.

The food is principally the seeds, fruits, tubers, roots, leaves or stems of many kinds of plants, shrubs, and trees. Many species eat more or less insects and a few eat flesh occasionally. Most species are strictly diurnal, but the Flying Squirrels are principally nocturnal. Some species are arboreal, some are terrestrial and many are fossorial.

The sexes are alike. The young differ but little from adults. In many species there are considerable seasonal changes of pelage, and a number of species are dichromatic, but dichromatism does, not appear to occur in any Californian species. There are usually four pairs of mammæ, occasionally five. The number of young in a litter varies greatly, rarely as few as one or two, frequently six or eight, rarely ten.

Genus **Marmota** FRISCH. (Marmot.)

Skull very short, broad posteriorly, narrow and flattened between the orbits; anteorbital foramen rather large, oval or pear shaped; postorbital processes long; penutimate premolar comparatively large; small internal cheek pouches; ears rather short; tail rather broad, about one third as long as head and body; inner toe on front foot rudimentary, with a flat nail; habit fossorial: mammæ ten; pelage coarse; size very large for the family.

Dental formula I, 1—1 ; C, 0—0 ; P, 2—1 ; M, 3—3×2=22.

Marmota flaviventer AUD. and BACH. (Yellow—belly.)
YELLOW-BELLIED MARMOT.

Above grizzled brown, the hairs being whitish at tip, with a broad pale chestnut zone and pale drab base; forehead, chin and lips dull white; nose sepia; top of head dark sepia; sides of neck buff; fore legs, hind feet and under surface of body varying from yellowish brown or wood brown to burnt umber; tail russet or cinnamon rufous on both surfaces.

Length about 485 mm. (19 inches); tail vertebræ 150 (6); hind foot 70 (2.75).

Type locality, Mountains between Texas and California.

Yellow-bellied Marmots inhabit the higher parts of the Sierra Nevada and the mountainous parts of northern California, the same or a closely related species being found in the mountain ranges north and east to the Rocky Mountains. They live in crevices of rocks near valleys and meadows in the higher mountains, and rarely burrow in level land. I have seen no evidence of their presence lower than 5,000 feet altitude. Their upper limit is unknown, but probably they go nearly to timber line in favorable localities.

The food is necessarily limited to those things of a vegetable nature that are available. In spring when but little bare ground was visible I found freshly cut juniper twigs about the entrance to their burrows. In summer a variety of succulent plants are to be had and then the Marmots take on fat rapidly. With the advent of freezing weather the Yellow-bellied Marmots probably bibernate, as the eastern species is well known to do.

Their hearing and sight are good. Their note is a single loud, clear whistle. Their breeding season is early, as I shot a suckling female in May above 7,000 feet altitude, where snow drifts were still deep.

Very few of these animals live near cultivated fields and hence they are practically harmless to man's interests. Marmots are often eaten, but to my taste their flesh is too rank to be agreeable. They are often called Wood-chuck and also Ground-hog.

Genus **Citellus** OKEN.

Skull varying in comparative width; postorbital processes usually well developed; penultimate premolar present, usually one-quarter to one-third as large as the last premolar; cheek pouches rather large; ears varying with species from large to rudimentary; tail varying in length and breadth; inner toe on front foot rudimentary; habit fossorial.

Subgenus **Otospermophilus**. (Ear—spermophile.)

Ears large; tail nearly as long as head and body, full haired; pelage mottled; audital bullæ rather small, with large and well rimmed external orifices; skull comparatively long and narrow.

Citellus beecheyi RICHARDSON. (For Captain F. W. Beechey.)

CALIFORNIA GROUND SQUIRREL.

Size large; tail long and comparatively bushy; ears large; back and sides thickly sprinkled with indistinct small whitish or pale brown spots on a sepia or drab ground, each spot bordered behind with dusky, the spots with a tendency to coalesce in irregular bars; a whitish patch on the sides of the neck, commencing behind the ears and prolonged across the shoulders in a stripe ending on the upper part of the side, these neck patches usually distinct and separated from each other by a pointed extension of the color of the back; top of head bistre grizzled with whitish; eyelids grayish buff or white; feet, sides of head and sometimes the face brownish gray; inner (concave) surface of ears and back border of outer surface yellowish gray, remainder of ears black; below brownish white or grayish; tail grizzled brown, the hairs having two or three dull black rings and the remainder, including base and tip yellowish white, the under surface of tail grayer than the upper side. *Young;* paler; white neck patches distinct; spots on sides and back dim.

Length about 415 mm. (16.33 inches); tail vertebræ 170 (6.70); hind foot 55 (2.15); ear from crown 20 (.80).

Type locality, California, probably Monterey, possibly San Diego.

California Ground Squirrels are abundant in nearly all parts of central and southern California, frequenting open valleys, brush and rocky hillsides alike; any sort of place that will supply abundant food will answer, but the borders of open ground where they can retreat to the cover of brush or rocks is preferred. They are found from sea level to the pine belt, to 8,000 feet altitude in southern California.

The food is principally of a vegetable nature, preferably grain and other seeds, fruit, potatoes, green plants, etc. Eggs of poultry and wild birds are relished. Some insects, such as grasshoppers are eaten, which is one of the too few things that can be put to their credit. On the label accompanying one of my Ground-Squirrel skins is the note "cocoons in the cheek pouches." Flesh is sometimes eaten. These Ground-Squirrels are serious pests to the farmer, and, in isolated places, to the fruit grower. They can climb well enough to get into peach and other fruit trees, and they make serious waste in small orchards in some places, as I know from personal experience. One often sees a strip several yards wide around a field stripped of the grain, and patches cut through the fields where they have established burrows.

There are various methods of destroying Ground-Squirrels, their effectiveness depending much on local conditions. In large areas of grain land poison is probably the most effective for·a beginning. In some places men make a business of poisoning the Ground Squirrels, taking contracts by the acre or for the ranch. Each man has some favorite method or formula for preparing the poison that he has been successful with. It will usually pay best to give the expert the job if one can be employed. The mistake is usually made of letting the work stop after the expert has filled his contract. If the area is large, so that when once cleared of Ground-Squirrels there is a good chance to keep it clear, every effort should be made to completely exterminate the pests. After the most thorough poisoning a few animals will be left, enough to re-stock the fields in a year or two, and if these are destroyed by one method or another before the next breeding season it will be comparatively easy to keep the borders clear in the future. Poisoned grain for destroying Ground-Squirrels is sold in all the towns, and unless large quantities are used it is as well to buy it ready prepared. Put a teaspoonful of the grain in each squirrel hole. In a few days repeat the dose in all holes that appear to be still used. Some Ground-Squirrels get cautious and do not eat enough to kill them. These may perhaps be disposed of by the use of bisulphide of carbon, which is fairly effective, and nearly as cheap as poisoned grain. The crude bisulphide is best as well as cheapest for this purpose. Its vapor is heavier than air and flows down the burrow, replacing air, and killing by suffocation. About a tablespoonful should be put on a bunch of rags or dry balls of horse manure and placed in each entrance to the burrow. It is best to close the mouth of the burrow. Coal oil fumes are successful if the burrow is not too large; use the coal oil in the same manner as the bisulphide of carbon. It is the cheapest of all methods but not always successful as the fumes are not as strong.

Filling the burrows with the smoke of straw and sulphur by using some of the patent smokers does well if the work is done thoroughly, but it is rather slow work on a large scale. Strychnine may be put on bits of apples, potatoes, etc., melon rind being particularly useful. Trapping with No. o steel traps is effective in a small place. The trap may be placed in the mouth of the burrow and lightly covered with dust or left bare. Grain or fresh meat may be used as bait, but if the trap is well located in the mouth of the burrow bait is scarcely necessary. Stake the traps well. Shooting may be the best method in some places. Often it is necessary to use one of these methods after another to get rid of the last squirrel, which may be exceptionally located, or unusually shrewd, but perserverance will conquer in the end unless ones place joins land that cannot be cleared.

The common note of the California Ground-Squirrel is a single loud whistle, short, repeated at intervals. When cornered in a rockpile or similar place they utter an angry chirring sound. Their sight and hearing are good. They do not hibernate, but in cold weather they remain in their burrows several days at a time, but a warm spell soon brings them out. The number of young at a birth is five to ten; they are born from the middle of April to the first or middle of June, according to locality.

Citellus beechey douglassi RICHARDSON.   (For David Douglass.)

DOUGLASS GROUND-SQUIRREL.

Pattern of coloration as in *beecheyi;* light spots white; gray tips of hairs of tail whiter; a wedge shaped black area on the shoulders and neck between the light neck patches; black stripe on ears indistinct, sometimes lacking; occipital rest of skull heavier. In examples from Mendocino and Lake Counties the hoary patches on the sides of the neck are nearly as dark as the sides. These western animals have smaller feet than those from the northeastern part of the State.

Length about 445 mm.   (17.50 inches); tail vertebræ 195 (7.66); hind foot 57 (2.25).

Type locality, Columbia River, eastern Oregon.

Douglass Ground-Squirrels are more or less common in the valleys in the northern part of the State, though not as abundant as the southern form often becomes. They are found in many parts of Oregon and in Washington. In the northern parts of the Sacramento Valley the Ground-Squirrels are intermediate between the Douglass and Californian subspecies. The habits of these races are similar.

Citellus beechey fisheri MERRIAM.   (For Dr. A. K. Fisher.)

FISHER GROUND-SQUIRREL.

Similar to *beecheyi* but everywhere paler; sides of neck and shoulder stripes clear silvery white, in strong contrast with the color of the body; sides of body thickly beset with indistinct whitish spots, narrowly bordered with dusky posteriorly; black ear stripe not sharply defined; eyelids and lower part of face whitish, under parts and feet buffy.

Size of *beccheyi.*

Type locality, 25 miles above Kernville, California.

Fisher Ground-Squirrels occur in the southern part of the

Sierra Nevada and eastward to the Panamint Mountains, but are common in few places. They prefer rocky hillsides bordering valleys. They are a desert race of the Californian Ground-Squirrel, with otherwise similar habits.

In the Providence Mountains, in the eastern part of the Mohave Desert, are a few Ground-Squirrels that I suppose are some form of *Citellus grammurus,* but I have no examples and cannot place them positively. As near as I remember the appearance of those that I sent the National Museum the whitish neck patches were indistinct and confluent and the hind parts were tinged with reddish brown. They are about the size, general appearance and habits of the California Ground-Squirrel.

Subgenus **Xerospermophilus**. (Dry—spermophile.)

Ears rudimentary; tail various in length and shape, usually narrow and one-fourth to one-half as long as head and body; pelage usually plain, sometimes striped; skull wide and strong; size small.

Citellus tereticaudus BAIRD. (Round-tail.)
ROUND-TAILED GROUND-SQUIRREL.

Winter pelage; above pale brownish cream buff; below creamy white; hairs comparatively long and soft. *Summer pelage;* above from nose to tail pinkish drab; below, sharply outlined along the sides, white; hairs short and coarse. *Young;* similar to winter adult.

Length about 240 mm. (9.50 inches) ; tail vertebræ 95 (3.75) ; hind foot 35 (1.40) ; ear a mere rim.

Type locality, old Fort Yuma, California.

Round-tailed Ground-Squirrels inhabit southeastern California, southern Arizona, northwestern Sonora and northeastern Lower California. In California they are most common in the lower Colorado Valley and in a few places in the Colorado Desert, though not really common anywhere.   Some occur in the Mojave Desert.   They avoid the rocky hills, preferring the level land.

The food is seeds the greater part of the year; these are stored to some extent.   In the spring, during the few weeks when green vegetation is obtainable, leaves and buds are eaten voraciously, the usually slender squirrel distending its stomach until it can hardly crawl away.

The voice is a peculiar low hissing whistle, sounding more like the note of some bird.   This note is uttered at intervals by the Squirrel when concealed in the mouth of its burrow, and is likely to puzzle one to account for it when first heard.

The breeding season is March and April.   The number of young in a litter is four to seven.   I have seen several Round-tailed Squirrels in low mesquit trees, where they were apparently feeding on the leaves, but they seemed awkward and slow climbers.   They are commonly shy and difficult to shoot.

Citellus beldingi MERRIAM.   (For Lyman Belding.)
BELDING GROUND-SQUIRREL.

A broad indeterminate band of chestnut or umber from nose to tail, more or less interrupted on the neck, varying in intensity from dark chestnut to dull raw umber; sides brownish or yellowish drab; below pale dull brownish or yellowish gray; tail small and slender, its upper surface similar to the back, lower surface cinnamon rufous or hazel edged with grayish and tipped with black. *Young;* dorsal stripe pale brown.

Length about 260 mm. (10.25 inches); tail vertebræ 70 (2.75); hind foot 42 (1.65); ear from crown 8 (.30).

Type locality, Donner, California.

Belding Ground-Squirrels are common in the valleys of the Sierra Nevada from the northwestern part of Inyo County north to the Oregon line.   In the western part of their range they do not seem to pass much below 5,000 feet altitude.   They reach to 9,000 feet in places.   In a few localities in Lassen and

Modoc Counties they do some harm, more particularly by reducing the pasturage where they are abundant.

The food is principally grass and annual plants, but on one occasion I caught a Belding Ground-Squirrel in a meat baited trap, and it is probable that they eat grasshoppers and other insects as well as seeds. They are not edible as their flesh is very rank. The alarm is half a dozen loud, clear, sharp whistles rapidly uttered. They often sit up very erect, with the fore-feet held close to the breast. This habit has given them the name of "Picket-pin", from the resemblance to a stake driven in the grass. They are also known locally as "Prairie-dog" and "Woodchuck."

The young are born late, as might be expected from the altitude of the region which they inhabit, from the latter part of May to July. This species hibernates regularly, going into winter quarters in September.

Citellus mollis stephensi MERRIAM. (Soft; for F. Stephens.)
STEPHENS GROUND-SQUIRREL.

Pelage comparatively long and soft; head, neck and shoulders grayish buff; back grizzled buffy drab; tail above like the back, the tip and under side grayish buff; below soiled buffy white.

Length about 212 mm. (8.33 inches); tail vertebræ 50 (2); hind foot 33 (1.30); ear from crown 3 (.12).

Type locality, Queen Station, north end of Owen Valley, Nevada.

The Stephens Ground-Squirrels were rather common in the valleys of eastern Mono County, California, and the adjoining part of Nevada, from 5,000 to 7,000 feet altitude. They were feeding on the sage brush and were excessively fat. Their habits appeared to be similar to those of the Mohave Ground-Squirrels.

Citellus mohavensis MERRIAM. (Of the Mohave Desert.)
MOHAVE GROUND-SQUIRREL.

Above uniform grizzled brownish drab or pinkish drab; upper side of tail similar to the back with more black intermixed; below dull buffy white.

Length about 230 mm. (9 inches); tail vertebræ 70 (2.75); hind foot 37 (1.45); ear rudimentary.

Type locality, Mohave River above Victorville, California.

The Mohave Ground-Squirrel may be distinguished from the Stephens Ground-Squirrel by its shorter and coarser pelage, longer and broader tail, darker head, and larger average size. In color it is usually darker than the Round-tailed Ground-Squirrel and the tail is shorter and broader, that of the latter being rat-like. The habits of all three species are similar. The Mohave Ground-Squirrel seems to be confined to the western and central part of the Mohave Desert. They do not appear to be very common.

Subgenus Callospermophilus. (Beautiful—spermophile.)

Ears rather large; tail about half as long as head and body, flat; pelage striped; nasals extending back further than premaxillaries; crown rather flat.

Citellus chrysodeirus MERRIAM. (Gilded.)
GILDED GROUND-SQUIRREL.

Pelage long and rather coarse, heavily striped; tail of medium breadth. *Summer pelage;* top of head chestnut; eyelids buffy; sides of head, neck and shoulders ochraceous or cinnamon; throat and legs dull ochraceous buff; broad dorsal band grizzled grayish brown, sometimes tinged with rufous, this band usually distinguishable to the crown and spreading over the rump and hips; two black stripes on each side, inclosing a buffy white stripe of about equal width which is usually tracable to ears and tail; sides and lower parts pale grayish ochraceous, the hairs of the belly dusky at base, this shade showing through; tail dusky above, edged with ochraceous, cinnamon beneath. *Winter pelage;* head and neck gray, more or less tinged with ochraceous. *Young;* similar to summer pelage; hairs long and coarse.

Length about 260 mm. (10.25 inches); tail vertebræ 93 (3.65); hind foot 40 (1.60); ear from crown 15 (.60).

Type locality, Fort Klamath, Oregon.

Gilded Ground-Squirrels are common in the Sierra Nevada and other mountains northward, in the pine timber, but are not often seen below 4,000 feet altitude. On the eastern slope they sometimes occur out of the pine timber. They prefer open forests with occasional rockpiles and old logs for places of refuge, under which they burrow. They feed on many kinds of seeds, on succulent plants, on mushrooms, and on the bulbs of such plants as have bulbs near the surface. I caught a number in traps baited with meat, and they probably eat various kinds of insects.

Five or six young are born between the middle of May and the first of July. As with others of this genus the mammæ are ten in number. They have the habit of standing erect. They are said to hibernate, which is no doubt true. I found them out in the high Sierras up to the middle of September, but they then appeared nearly ready to go into winter quarters.

Citellus chrysodeirus bernardinus NELSON. (Of the San Bernardino Mountains.)

SAN BERNARDINO GROUND-SQUIRREL.

Very similar to *chrysodeirus;* "tail and hind foot shorter; duller mantle over head and shoulders."

Type locality, San Bernardino Peak, California.

Rather common about Bear Valley, San Bernardino Mountains and occasional in other parts of that range. I have not teen this species in any other range in southern California, though both the San Jacinto and San Gabriel Mountain ranges are well adapted to their wants. *Bernardinus* differs very little from true *chrysodeirus.*

Citellus chrysodeirus trinitatus MERRIAM. (Of the Trinity Mountains.)
TRINITY GROUND-SQUIRREL.

In fall pelage similar to *chrysodeirus;* larger; ground color darker; inside of tail dark chestnut; skull and teeth larger; nasals longer.

Length about 280 mm. (11 inches); tail vertebræ 100 (4); hind foot 43 (1.70).

Type locality, Trinity Mountains, California.

"Common in the Siskiyou, Salmon and Trinity Mountains of northwestern California and southwestern Oregon." I have not seen this subspecies. It is supposed to lack the golden mantle of the head and shoulders.

Subgenus **Ammospermophilus**. (Sand—spermophile.)

Ears small; tail about half as long as head and body; pelage striped; nasals small and short; rostral part of face small: crown well arched; skull narrow and light; size small.

Citellus leucurus MERRIAM. (White—tail.)
ANTELOPE GROUND-SQUIRREL.

Above smoke gray or drab grizzled with white; a narrow white stripe from shoulder to hip, below which is a broad stripe on the side similar in color to the back, but often tinged with cinnamon; outer surface of legs vinaceous cinnamon or vinaceous buff; eyelids, inner surface of legs and lower parts white; tail short, broad, flat, white underneath, mixed black and white above, the hairs being black at root, then white, then black and tipped with white; tail often tinged with salmon above at base.

Length about 210 mm. (8.25 inches); tail vertebræ 66 (2.60); hind foot 37 (1.45); ear from crown 6 (.25).

Type locality, San Gorgonio Pass, below Banning, California.

Antelope Ground-Squirrels are more or less common in the hills bordering the Colorado and Mohave Deserts and in rocky places in these Deserts and north to Lassen County. In a very few places they occur a short distance down on the Pacific slope. They are not often seen out on open plains, preferring rocky localities.

The food is principally seeds as is usual with Ground-Squirrels. The cheek pouches together will hold more than a heaping teaspoonful of seeds. The note is a loud, prolonged, tremulous whistle. The breeding season is early, March and April. Five to eight is the usual number of young. In running these Ground-Squirrels carry their tails curled over their backs, the underside of the tail appearing like a white rump, hence their common name.

Citellus nelsoni Merriam. (For E. W. Nelson.)
NELSON GROUND-SQUIRREL.

Similar to *leucurus* but larger and paler; above dull yellowish brown or buffy clay color, dark beneath the surface; white lateral stripe tinged with ochraceous; outer surface of legs and upper side of tail near base, buffy clay color; remainder of upper side of tail black and white, the white border broad; lower part buffy white. In winter the back is nearly as dark as *leucurus*.

Length about 223 mm. (8.65 inches); tail vertebræ 70 (2.75); hind foot 40 (1.60).

Type locality, Tipton, San Joaquin County, California.

The Nelson Ground-Squirrel is found in the southern part of the San Joaquin Valley, where it is common in a few places.

Judging from my limited experience with this species they occur in open ground principally. I saw none in rocky places. Those I saw were very shy.

Genus **Eutamias** TROUESSART. (Good or typical; a steward.)

Skull light and thin, moderately arched in upper outline; postorbital processes small and slender; penultimate premolar present but small, rarely functional; anteorbital foramen small, oval, with a prominent tubercle at its lower edge; internal cheek pouches large; ears rather large; tail about as long as body without the head; inner toe on front foot rudimentary; pelage striped, rather long; mammæ eight; habit terrestrial and fossorial; size small.

Dental formula, I, 1—1 ; C. 0—0 ; P, 2—1 ; M, 3—3×2=22.

The species of this genus are very much alike, and even experts are sometimes in doubt what species a given specimen should be assigned to when material for comparison is not available. The skulls show no differences of sufficient value to be of much use in separating species, and there is considerable individual variation in externals. More or less marked seasonal variations in pelage help complicate the situation. The group is very "plastic," being usually susceptible to modification from climatic causes, in the directions of size and color; hence differences in climatic through differences of altitude, isolation on mountain ranges separated from other mountains by climatic barriers, etc., have brought about a separation of the genus into a number of closely related forms. From the nature of these causes the lines of separation are often indistinct.

The following characters are common to all Californian species in fresh pelage, in old worn pelage some points will be lacking:

Five dark stripes on the back from the neck or shoulders to the rump, enclosing four light stripes in decided contrast;

three dark and two light stripes on each side of the face, some of
these indistinct in some forms; top of head from nose to ears
gray or brown; ear striped, the front edge usually reddish, the
front half of the outer (convex) surface black or blackish, the
back half gray or white, usually in strong contrast; a gray or
white spot behind each ear, varying in distinctness; lower surface of body light colored, white, gray or buffy; sides gray, more
or less tinged with buff or rust color, this tinge strongest in summer pelage; feet grayish or brownish; hairs of upper surface of
tail and its edges banded in contrasting colors, each hair being
blackish at base, then buffy or rusty, then black, then with a
longer or shorter tip of buff, gray or white; hairs of under surface of tail dusky at base, the remainder yellowish or reddish,
producing a distinct light stripe the length of the tail, except
tip which is black.

The summer moult takes place at the end of the breeding
season, earlier with males. The summer pelage is brighter than
that of winter and lasts about three months. In the spring the
colored tips to the hairs on the back may be so worn that the
stripes are indistinct or entirely worn off.

Eutamias alpinus MERRIAM. (Alpine.)
ALPINE CHIPMUNK.

Very small and very pale in winter pelage; dark face stripes
narrow, pale rusty brown; top of head gray; ear markings pale;
spot behind the ear dull white, diffused; outer pair of light stripes
broad, white, inner pair pale gray or grayish white, middle back
stripe rusty brown the others fulvous, more rust colored in summer and the outer pair shading into the fulvous of the sides,
which, are grayer in spring; tail rather long, broad, ochraceous
buff below, very narrowly bordered with black and broadly edged
with buff, above blackish shaded with yellowish or hoary, black
toward the tip, particularly below.

Length about 185 mm. (7.30 inches); tail vertebræ 79
(3.10); hind foot 29 (1.15).

Type locality, Big Cottonwood Meadows, southeast of Mount Whitney, Cal.

Alpine Chipmunks are found high in the southern Sierras, living in the crevices of the rocks about timber line, going occasionally nearly to the summits of the highest peaks. They do not occur much below ten thousand feet altitude. I found them shy and hard to get, as when shot they were likely to fall in deep inaccessible crevices among the rocks.

Eutamias amœnus ALLEN. (Pleasant.)
KLAMATH CHIPMUNK.

Small; facial stripes distinct; spot behind the ear grayish white, not in strong contrast with surrounding darker pelage; outer light stripes narrow, grayish white; inner light stripes pale brownish gray; dark stripes of body, blackish, more or less edged or tinged with reddish; tail narrow, blackish, the hairs tipped with buffy white, the concealed band pale yellowish much lighter than the tawny olive or ochraceous buff stripe on the under side; sides russet, more grayish in winter pelage, and shading to the buffy white belly.

Length about 198 mm. (7.80 inches); tail vertebræ 90 (3.55); hind foot 31 (1.22); ear from crown 16 (.63). Weight about two and one-half ounces.

Type locality, Fort Klamath, Oregon.

Klamath Chipmunks are common or abundant in the high mountains from Inyo County north to Idaho and Washington. Their range in the Sierra Nevada is from about 4,000 to 8,000 feet altitude. They frequent the brushy places and the open timber adjoining, but are rarely seen in thick forest. Three to five young are born in May and June.

Eutamias pictus ALLEN. (Painted.)
DESERT CHIPMUNK.

Similar to *amœnus;* tail narrower and concealed stripe on upper side of tail about the color of that on the under side;

dark stripes on back black or brownish black slightly edged with chestnut and contrasting sharply with the light stripes; sides pale buffy gray in winter, tawny ochraceous in summer; specimens from the middle and eastern parts of their range are probably paler.

Length about 195 mm. (7.70 inches); tail vertebræ 93 (3.65); hind foot 30 (1.18).

Type locality, Kelton, Utah.

Desert Chipmunks are rather common in the sage brush plains of the Great Basin, having their western limit in the eastern foothills of the Sierra Nevada, from the Owen Valley north to Oregon. Four to six young are born in May and June. I saw these Chipmunks occasionally in juniper trees and they frequently climb to the tops of the sagebrush.

Eutamias panamintus MERRIAM. (Of the Panamint Mountains.)
PANAMINT CHIPMUNK.

Small; dark facial stripes indistinct; ear markings obscure; dark back stripes fulvous brown and not reaching the rump which is clear gray; sides gray washed with buffy ochraceous in winter pelage, more rusty in summer; upper side of tail orange rufous thinly shaded with black and washed with yellowish.

Length about 208 mm. (8.20 inches); tail vertebræ 90 (3.55); hind foot 31 (1.22).

Type locality, Panamint Mountains, California.

Panamint Chipmunks inhabit the pinon and pine timber of the Panamint and other isolated desert ranges. They are common in few places.

Eutamias speciosus ALLEN. (Appearing well.)
SAN BERNARDINO CHIPMUNK.

Size medium; facial stripes very distinct, that one passing across the eye broad and black; ears distinctly striped; spots be-

hind the ear large, white and distinct in summer, less distinct in
winter; in summer the dark stripes of the back are prout brown,
more or less tinged with rusty, the outer pair shading into the
burnt umber of the sides; in winter all these stripes and the sides
are grayer and duller, middle pair of light stripes brownish gray,
the outer pair broad and white; lower parts white shading into
the gray of the lower part of the sides; tail rather short and
narrow, russet or cinnamon rufous below, indistinctly bordered
and broadly tipped with black, edged with yellow; tail above
rusty black, the russet band showing through more or less. In
badly worn spring pelage the stripes on the back may be nearly
or quite obliterated, leaving the back plain.

Length 210 mm. (8.25 inches); tail vertebræ 95 (3.75);
hind foot 33 (1.40); ear from crown 16 (.63).

Type locality, San Bernardino Mountains, California.

San Bernardino Chipmunks are common in the higher parts
of the San Bernardino and San Jacinto Mountains. They are
better climbers than most other species of this genus, frequently
running up the smaller trees and sitting on a knot or limb and
chipping at the passer-by. The breeding season is May and June.
The number of young appears to be four and five.

Eutamias speciosus callipeplus Merriam. (Beautiful— mantle.)
MOUNT PINOS CHIPMUNK.

Similar to *speciosus;* dark dorsal stripes narrower; outer
white stripe broader and white spot behind the ear larger; in
summer pelage the sides are cinnamon rufous and the rump is
tinged with rusty; in winter the sides are gray with a rusty tinge
and the back stripes are grayer.

Type locality, Mount Pinos, Ventura County, California.

This subspecies appears to be limited to the small area of
Mount Pinos at the extreme southern end of the San Joaquin
Valley and to the upper part of the western slope of the southern

Mt Pinos Chipmunk. One-third life size.

Sierra Nevada. I have not seen examples from the western side of the southern Sierras, it would seem that these should be nearer *frater* as there is no break in this part of the range as there is to the southward.

Eutamias speciosus inyoensis Merriam. (Of Inyo County.)

INYO CHIPMUNK.

Similar to *speciosus;* larger; facial stripes less distinct; light spot behind the ear indistinct; back of neck gray; middle stripe of back mostly black; rump grizzled golden yellowish; upper side of tail very fulvous; black tip of tail short.

Type locality, White Mountains, Inyo County, California.

The Inyo Chipmunks are found in the higher parts of the White and Inyo Mountains along the eastern border of California. They are not common.

Eutamias speciosus frater ALLEN. (Brother.)
SIERRA NEVADA CHIPMUNK.

Rather larger than *speciosus;* facial stripes not so prominent; outer light stripes of back narrower, white tinged with buff; dark stripes rather broad, seal brown or dark chestnut; belly grayish white; sides very red in summer pelage, between russet and cinnamon rufous, winter pelage grayer.

Type locality, Donner, California.

Sierra Nevada Chipmunks range through the Sierra Nevada, excepting perhaps the southwestern part, from 5,000 to 9,000 feet altitude, coming lowest in the northern part of their range; where they occur in the yellow pines. They climb trees to some extent. They do not seem to be very shy.

Eutamias quadrimaculatus GRAY. (Four—spotted.)
LONG-EARED CHIPMUNK.

Size large; facial stripes very distinct; ears larger than any other Californian species of the genus; front edge of convex side of the ear rusty brown shading into blackish sharply bordered behind with white, the contrast less sharp in summer; dark stripes of back seal brown edged or tinged with rusty; outer light pair of stripes white and distinct in summer, grayer in winter pelage; sides fulvous gray in spring and fall, rusty brown in summer; lower parts white; tail long, of medium breadth, the stripe underneath rich light chestnut; tips of hairs of upper side of tail whitish.

Length about 235 mm. (9.25 inches); tail vertebræ 108 (4.25); hind foot 35.5 (1.40); ear from crown 20 (.78). Weight three to three and a half ounces.

· Type locality, Michigan Bluff, Placer County, California.

Long-eared Chipmunks are common in the yellow pine forests of the western slope of the Sierra Nevada from the Yosemite Valley north to Lassen County. I obtained a number of Chipmunks on the headwaters of the Carson River in Alpine County, that I suppose to be Long-eared.

These Chipmunks are found in thicker forests than most of the genus like. They climb trees but little. The young are born from the latter part of May until in July.

Eutamias quadrimaculatus senex ALLEN. (Old.)
ALLEN CHIPMUNK.

Similar to *quadrimaculatus;* ears considerably shorter and less pointed; facial stripes not as distinct; light spots behind the ear smaller and tinged with ashy; white markings grayer; size about the same.

Type locality, Donner Pass, Placer County, California.

Allen Chipmunks range from the higher peaks above the Yosemite Valley north along the Sierra Nevada and Cascade Mountains to central Oregon and west on the Siskiyou Mountains. I obtained Chipmunks in the Warner Mountains which I refer to this subspecies. Allen Chipmunks seem to be the northern and high mountain form of *quadrimaculatus.*

Eutamias hindsi GRAY.
HINDS CHIPMUNK.

Light face stripes grayish white in winter, tinged with ochraceous in summer; whitish ear stripe and spot behind the ear distinct, most so in summer; other dark stripes of back indistinct, the three middle stripes black, edged with rusty in winter; inner pair of light stripes wood brown or russet grizzled with gray in winter pelage, deeper brown in summer; outer pair of light stripes rather narrow, grayish white in winter pelage tinged with fulvous in summer; sides light burnt umber in summer, gray tinged with umber in winter; lower parts dull white in winter, washed with fulvous in summer; tail long, broad, the stripe underneath cinnamon rufous or light chestnut, rather narrowly bordered with black, edged with grayish white.

Length about 250 mm. (9.85 inches); tail vertebræ 118 (4.65); hind foot 36 (1.42).

Type locality, near San Francisco, California.

, The northern part of the range of the Hinds Chipmunk lies east of and joining that of the Redwood Chipmunk from eastern Humboldt County southward, and reaches the coast a little north of San Francisco. They prefer localities where mixed timber is interspersed with brush and patches of open grass land. The breeding season commences early, the latter part of March, the greater number being born in April. The young number from three to five.

Eutamias hindsi pricei ALLEN. (For W. W. Price.)
PRICE CHIPMUNK.

Winter pelage, similar to *hindsi* but tinged more with rusty and pelage more grizzled with white; apparently averaging larger. I have not seen the summer pelage.

Type locality, Portola, San Mateo County, California.

Price Chipmunks occur through the Santa Cruz Mountains and probably in the mountains of Monterey County also.

Eutamias merriami ALLEN. (For Dr. C. Hart Merriam.)
MERRIAM CHIPMUNK.

Most like *hindsi* but grayer. Light face stripes dull white, sometimes slightly tinged with ochraceous in summer; stripe on back part of ear gray, indistinct; spot behind the ear grayish white in summer, ashy gray in winter; dark stripes of back in winter dull brownish black, the middle stripe darkest and the outermost pair almost obsolete; ground color of upper surface and sides hair brown grizzled with whitish, particularly on the innermost pair of light stripes; sides tinged with raw umber; outer pair of light stripes, dull grayish white; belly grayish white sometimes tinged with buff; stripe on under side of tail cinnamon rufous, rather broadly bordered with black and edged with yellowish white. In *summer pelage* the sides are heavily tinged with russet or

fulvous and the dark stripes of back more or less tinged with the
same color; outer pair of light stripes whiter.

Length about 250 mm. (9.85 inches); tail vertebræ 120
(4.70); hind foot 36 (1.42); ear from crown 16 (.63).

Type locality, San Bernardino Mountains, California.

The Merriam Chipmunk is common in the San Bernardino
and San Jacinto Mountains, in the higher parts of the mountains
of San Diego County and on the lower western slope of the
southern Sierra Nevada. They do not extend very far below the
belt of coniferous trees, and are usually found in brush, rarely
climbing trees. Four or five young are born in the last half
of May or in June.

Eutamias townsendi ochrogenys Merriam. (For J. K. Townsend; pale yellow—chin.)

REDWOOD CHIPMUNK.

Large and very dark colored. Light face stripes pale
ochraceous, increasing in intensity as the season advances; back
part of ear and light spot behind the ear bluish white, distinct;
middle back stripe black, the next dark pair rusty black, outer
pair narrow and indistinct; inner pair of light stripes sepia
or bistre grizzled with white, outer pair pale olive gray; top of
head and rump grizzled dark sepia; sides bistre or wood brown,
lower parts ochraceous white; tail long and narrow, blackish
above, the stripe beneath chestnut broadly bordered with black
and edged with grayish white. *Summer pelage;* strongly tinged
with fulvous.

Length about 260 mm. (10.25 inches); tail vertebræ 117
(4.60); hind foot 37 (1.45); ear from crown 16 (.63).

Type locality, Mendocino, California.

Redwood Chipmunks are more or less common in the redwood forests in a narrow belt a few miles wide along the coast
of northern California from Sonoma County to Oregon. They
frequent the thick forest, seldom coming out in the open places.

They are fond of running along fallen trees, but I saw none climb standing trees. The number of young is unusually small, two and three in a litter. These are born in the latter part of May and in June.

Genus **Sciurus** Linneus. (Shade—tail.)

Skull short, broad between the orbits; anteorbital foramen very small; postorbital processes long, slender, bent obliquely backward and downward; penultimate premolar very small or absent; incisors narrow; inner toe on front foot very rudimentary; ears large, sometimes tufted; tail long and bushy; no internal cheek pouches; mammæ four to eight; diurnal; arboreal.

Dental formula, I, 1—1; C, 0—0; 2—1 or 1—1; M, 3—3×2=22 or 20.

Subgenus **Hesperosciurus**.

Skull comparatively long, strongly arched in upper outline; posterior part of cranium depressed; rostrum long and deep; nasals long and narrow.

Sciurus griseus Ord. (Gray.)
COLUMBIA GRAY SQUIRREL.

Size large; general color of upper surface of head, body and tail mouse gray thickly grizzled with white; eye ring dull white; ears in winter pelage clothed on the convex surface with soft fur, dusky at tip, light brown at base, in summer scantily haired, never tufted; no lateral stripe; under surface from chin to tail white to roots of hairs, sharply defined against the ashen gray sides; tail very large, long, bushy, flat, the hairs often three inches long, slate gray mixed with whitish annulations, each hair with a long white tip; under surface of tail pale ashy gray centrally with blackish lateral bands and white border; ashy gray of shoulders and hips extending down on the outside of the

ANTHONY GRAY SQUIRREL

legs and feet to the toes, becoming paler gray on the feet. Examples from northern California and the Sierra Nevada aᴵᴸ lighter colored than those from Oregon through greater exten ᴸ of the white grizzle on the upper surface.

Length about 560 mm. (22 inches); tail vertebræ 2ᴼᴼ (11); hind foot 79 (3.10); ear from crown 30 (1.18).

Type locality, near the Columbia River in Oregon or Wasnington. (From Lewis and Clark's description.)

Columbia Gray-Squirrels occur in western Wasnington, western Oregon, northern California and through the Sierra Nevada. This seems to be the form along the coast north of San Francisco. I have taken examples in Mendocino and Plumas Counties and in the southern end of the Sierras that vary no more than individuals do in any of these places. They are all grayer than those from Oregon, but not greatly so.

Sciurus griseus nigripes BRYANT. (Black—foot.)
BLACK-FOOTED GRAY-SQUIRREL.

Considerably darker than *griseus,* the white surface grizzling being reduced in amount and the sub-color darker; tail darker; upper surface of feet much darker, slaty or black, the hind feet often black to the toes; back more or less suffused with umber brown.

Black-footed Gray-Squirrels occur from San Francisco southward in the Santa Cruz Mountains. I have seen no examples from the Mountains of Monterey County, but probably they would be nearest this subspecies.

Sciurus griseus anthonyi MEARNS. (For A. W. Anthony.)
ANTHONY GRAY-SQUIRREL.

Very similar to *griseus;* feet darker, intermediate in color between *griseus* and *nigripes.* In winter the whitish grizzling becomes more ochraceous on the back, resulting in an indefinite brown band.

Length about 550 mm. (21.65 inches); tail vertebræ 207 (10.30); hind foot 76 (3); ear from crown 30 (1.18).

Type locality, Laguna Mountain, San Diego County, California.

Anthony Gray-Squirrels inhabit the higher mountains of southern California. They are rarely seen below the lower edge of the pine forests, about 4,000 feet altitude and prefer those localities where oaks are mixed through the conifers. They are not found much above the upper limits of the oaks, about 8,500 feet altitude.

The food is principally the seeds of coniferous trees and acorns, but other kinds of seeds are also eaten. The "bark" is a series of hoarse notes rapidly uttered, which can be heard a considerable distance. The hearing and sight are keen.

The breeding season is prolonged, the young being born from April to August. I do not believe that more than one litter is reared annually, as the young are born in a very immature condition, and remain in the nests a considerable time. When born they are blind and almost hairless. The number of young in a litter is one to four, two and three being the most frequent numbers. The breeding nests are large globular masses of twigs and leaves situated well up in trees. The ordinary dwellings are hollows of trees lined with leaves and strips of bark.

Anthony Gray-Squirrels do not hibernate, but in stormy weather they may remain in their nests several days at a time. In fine weather in winter they run about on the snow or bare ground, as the case may be, foraging for food. My impression is that they do not store up much food for winter use. ,

These Squirrels run about on the ground a great deal, ordinarily preferring to travel on the ground rather than through the treetops. They seem to be subject to epidemics and irregular fluctuations in abundance. They do practically no harm to crops and are fair eating.

Subgenus **Tamiasciurus**.

Skull short, moderately arched; posterior part of cranium wide, not greatly depressed; rostrum short; nasals wide and very short.

Sciurus douglassi albolimbatus ALLEN. (For David Douglass; white—border.)
CALIFORNIA CHICKAREE.

Summer pelage; above from crown to tail brownish gray tinged with tawny, the hairs being slightly tipped with this color; eye ring buff; ears large, more or less tufted with black. hairs; a black stripe on the side varying in length and distinctness with season and individual; fore legs and feet and hind feet tawny olive; under side of head and neck white, tinged with buff; remainder of under parts buff or ochraceous buff varying in intensity of color, the hairs being of this color nearly or quite to the roots; tail blackish mixed with ochraceous above and grayish below, the hairs tipped with white, most distinctly at the sides; terminal fourth of tail mostly black. *Winter pelage;* lower parts nearly pure white; tawny tips of hairs of upper parts longer; black stripe on side obscure; ear tufts longer.

Length about 330 mm. (13 inches); tail vertebræ 135 (5.25); hind foot 52 (2); ear from crown 21 (.83). Weight ten ounces.

Type locality, Blue Canon, Placer County, California.

California Chickarees are common locally in the coniferous forests of the Sierra Nevada and other mountains of the northeastern part of California, from 3,000 feet altitude up nearly to timber line, but are most common in mixed pine and fir forests. Not known to occur south of the Sierra Nevada.

The food is quite varied but consists principally of the seeds of conifers, such as fir, big trees, sugar pine and yellow pine, berries, nuts, acorns and chinquapins. Mushrooms and some insects are also eaten. I have caught these Chickarees in meat baited traps set for other animals. They often store seeds in shallow holes scratched in the ground. A portion of these seeds are not recovered and some germinate, so the Chickaree is quite an important agent in nature's system of tree planting. They usually bite off the stems of the large cones letting the cones fall to the ground, coming down to gnaw their seeds loose at the foot of the tree.

Chickarees are very active, keeping mostly in the larger trees, often running up and down their trunks, apparently for sport. They are wary, yet inquisitive, and if one keeps quiet they will soon come out on a knot to scold the intruder. The voice is varied, commonly a rapid chirring series of notes is heard, sometimes a sharp yelp, or again a bird-like note. Most of these sounds are emphasized by jerks and wags of the tail.

Chickarees do not hibernate in California, though in stormy weather they remain several days at a time in their nests in hollow trees, but in fine weather they run about on the snow as if they enjoyed the cold weather, as they doubtless do. They do not like hot weather and are not found in the warm valleys or low mountains.   ,

The breeding season is late, as the young are born in June, and July. The young are four or five in number. I have shot females in August that were then suckling young. The summer moult takes place in June and the autumn moult in September.

Sciurus douglassi mollipilosus Aud and Bach. (Soft— haired.)
REDWOOD CHICKAREE.

Similar to *albolimbatus;* color above darker ; below ochraceous buff or pale salmon shaded with dusky, the basal half of the hairs being dusky and showing through the tips. Averaging smaller.

Type locality, Coast of northern California.

The Redwood Chickaree is found principally in the redwood forests, more particularly where the firs are mixed among the redwoods, and in the oaks in the openings among the redwoods. I did not hear the chirring song that is made by most other subspecies, but heard the "bark," a monotonous "quoo" uttered at intervals of two or three seconds. When startled or disturbed this changed to a querulous "queeo." My impression from my brief acquaintance with the Redwood Chipmunk is that they are tame and unsuspicious. I saw them occasionally on the ground.

Genus **Sciuropterus** Cuvier. (Squirrel—wing.)

Upper outline of skull strongly arched; penultimate premolar present; anteorbital foramen triangular, rather small; skin of sides loose, extensible between the fore and hind legs to form parachute-like "wings" and extended by a long slender bone articulated with the carpus and directed backward and outward; tail long, broad, very much flattened; no cheek pouches; eyes large; ears of moderate size, thinly haired; pelage soft; size small; habit crepuscular and nocturnal; eight mammæ.

Sciuropterus alpinus klamathensis Merriam.
KLAMATH FLYING SQUIRREL.

Above dark drab brown, sometimes tinged with pale dull fulvous brown; under parts pale yellowish buff, the plumbeous under fur showing through; upper surface of tail like back, but somewhat darker, especially toward the end; under side of tail uniform deep buff; nose and feet pale; cheeks pale yellowish gray.

Length of type 329 mm. (13 inches); tail vertebræ 135 (5.45); hind foot 38 (1.50).

Type locality, Fort Klamath, Oregon.

Southern Oregon and probably northeastern California.

Dr. Merriam saw a Flying Squirrel on Mount Shasta which he thought was of this subspecies.

Sciuropterus alpinus californicus Rhoads.
SAN BERNARDINO FLYING SQUIRREL.

Similiar to *klamathensis;* apparently paler and smaller.

Length of type 286 mm. (11.25 inches); tail vertebræ 127 (5.10); hind foot 38 (1.50). A female that I took in the type locality measured, length 245 (11.60); tail vertebræ 140 (5.50); hind foot 37 (1.45); ear from crown 20 (.78). Weight five ounces.

Type locality, San Bernardino Mountains, California.

Flying Squirrels appear to be rare in southern California and are not known to occur below the coniferous forests. They are nocturnal in habit and may be more common than we suppose.

The food of Flying Squirrels consists of seeds, buds, beetles and flesh, occasionally at least; whether or not they habitually kill small mammals and birds is not certainly known. They live in holes in trees and rarely come out until twilight.

The flight of Flying Squirrels is not true flying but is a sailing leap. They leap from the upper part of a tree with the side membranes extended and with the aid of these and the broad flat tail sail down and out, alighting against the lower part of another tree, running up, to again leap from the top. They can guide the flight to some degree, but cannot rise to the height from which they started.

Sciuropterus oregonensis stephensi MERRIAM.
STEPHENS FLYING SQUIRREL.

Above wood brown, the tips only of the hairs being of this color, the remainder slate gray, this color showing through the tips; upper part of head and neck a lighter brown; a narrow blackish eye ring; sides of head and cheeks pale brownish gray; feet drab gray; under surface of head, body and wings white tinged with pale brownish yellow, the slaty under fur showing through; upper surface of tail mouse gray tinged with drab toward the base; under side of tail light smoke gray, darker at the edges.

Length of type 277 mm. (10.90 inches); tail vertebræ 131 (5.15); hind foot 37 (1.45); ear from crown 19 (.75).

Type locality, Sherwoods, Mendocino County, California.

I caught the type of this subspecies in a thick redwood forest, in a steel trap baited with meat and set for mink at the roots of a large redwood tree a few inches from a brook.

Family. **Aplodontidæ.** Sewellels.

Skull massive, flat, much constricted interorbitally, excessively widened posteriorly; brain case comparatively small; zygomatic arches widened posteriorly; no postorbital processes; anteorbital foramen small, low, oval; nasals short and broad; audital bullæ peculiar, tubular, being greatly lengthened laterally; descending ramus of lower jaw very wide with a projecting lateral angular flange; coronoid process high; molariform teeth simple, roctless, prismatic, penultimate upper premolar present but small; five toes on each foot, the inner toe of front foot small but functional; tibia and fibula separate though closely apposed; outlets of genito-urinal and digestive organs separate.

This peculiar family contains but a single genus, consisting of half a dozen species and subspecies. It appears to be one of the most primitive types of mammals now existing, having no very close affinities with any other living family. It is of limited distribution, being found only in western North America from California to British Columbia, in the Sierra Nevada and Cascade Mountains and in parts of the lower region west to the Pacific coast.

The food is twigs, stems and leaves of shrubs and plants. mostly perennial. They are plantigrade, nocturnal, semi-aquatic, fossorial, living in burrows in wet ground. The sexes are alike; the young are darker in color but are otherwise similar to the adults. There are five pairs of mammæ, nearly equally distributed from the armpit to the groin.

Genus **Aplodontia** Richardson. (Simple-tooth.)

Eyes small; ears projecting a short distance above the surrounding fur; no cheek pouches; neck short and thick; legs short; claws of fore feet largest; feet not webbed; soles naked; tail haired and very short; form depressed, stout; pelage consisting of thick underfur mixed with long hairs.

Dental formula, I. 1—1; C, 0—0; P. 2—1; M, 3—3×2=22.

Aplodontia major MERRIAM. (Greater.)
CALIFORNIA MOUNTAIN BEAVER.

Above from nose to hips, and on the sides grayish sepia brown grizzled with black, the pelage slate colored at base and the long intermixed hairs black tipped; hips, rump, tail and under parts smoke gray; a small white anal spot; whiskers mostly black. *Young;* slate brown.

Length about 355 mm. (14 inches); tail vertebræ 42 (1.60); hind foot 62 (2.45); ear from crown 8 (.32). Weight three to four pounds.

Type locality, Placer County, California.

California Mountain Beavers occur in isolated localities in the Sierra Nevada and northward, and also in the Siskiyou Mountains. I have taken them in Alpine County, on the eastern slope at the headwaters of Carson River.

They live in wet springy land in canyons and on mountain sides where suitable springs occur, usually at considerable altitudes. I obtained mine at 8,000 feet altitude. The burrows in most cases ran up and down the wet hillside, for drainage, and often had openings every few feet. Some of the burrows were fifty yards or more in length, and in a few cases spring brooks had broken into the upper entrance and ran in the burrows instead of in their natural channels. In one case a brook was diverted from its own channel to that of one several yards away. Most of the entrances to the burrows were under clumps of willows. Many of the burrows had more or less water running from their lower entrances, but rather the greater number were dry.

The plants that I saw cut for food were an *Iris,* an *Astragulus,* willow and alder. To these Allen adds fir, manzanita and lilies; and Price *Ceanothus, Rhododendron* and mountain cranberry. Probably many other plants are also eaten. They can climb bushes, and Allen and I each saw brush and small trees trimmed off three or four feet from the ground. I saw bunches of plants laid up on low bushes to dry, commonly over entrances to burrows, most of these not being much dried, as if they carried them in as soon as they were well wilted.

All the animals caught were alive when I reached them in the morning. None had made any attempt to gnaw off the leg, as true beaver would have done. Most of them were the length of the trap chain down their burrows. While pulling them out they made a whining sound. Some showed fight. They used their hind feet in grasping as readily as their fore feet and as well as a squirrel. It would appear that other animals prey on the Mountain Beaver as I caught a weasel and two skunks in traps set for Mountain Beaver. Hibernation is probably imperfect. The fur is of no value.

Aplodontia phæa MERRIAM. (Dusky.)
POINT REYES MOUNTAIN BEAVER.

Similar to *major;* smaller; above grizzled bistre.

Length of type specimen 330 mm. (13 inches); tail vertebræ 30 (1.20); hind foot 55 (2.15).

Type locality, Point Reyes, Marin County, California. Limits of distribution unknown. The only other record that I have seen that may apply to this species is that of a specimen in the collection of the California Academy of Sciences from near Eureka.

Family **Castoridæ.** Beavers.

Skull massive, flat, not constricted interorbitally, nor excessively widened posteriorly; zygomatic arches widened posteriorly; no postorbital processes; nasals short, broad, oval in outer outline; audital bullæ moderately lengthened laterally; descending ramus of lower jaw wide but of normal shape; molariform teeth single rooted, with the pulp persisting late in life; planes of upper molars convergent anteriorly; outlets of genito-urinal and digestive organs combined in one.

Dental formula, I, 1—1; C, 0—0; P, 1—1; M, 3—3×2=20.

The Beavers are a very small family, containing but one living genus, consisting of but two species as now recognized. They are distributed over the colder parts of the northern hemisphere. The food is strictly vegetable, consisting mostly of twigs and bark obtained by gnawing down trees and shrubs. Their fur is valuable and has been an important article of commerce.

Beavers are plantigrade, nocturnal, semiaquatic and live in burrows or in "houses" constructed of sticks and mud. The males are somewhat larger than the females but the sexes are otherwise alike and the young differ but little from the adults.

Genus **Castor** LINNEUS. (Beaver.)

Form stout; tail broad, flat, tongue shaped, covered with scales instead of with hairs; front feet small, not webbed, the inner toe developed but smaller than the others; no cheek pouches; pelage consisting of thick fine underfur interspersed with long coarse hairs.

Castor canadensis frondator MEARNS. (Twig-stripper.)
BROAD-TAILED BEAVER.

Above russet; below grayish cinnamon; sides wood brown; feet burnt sienna color.

Length of adult male about 1090 mm. (43 inches); tail

vertebræ 355 (14); hind foot 185 (7.25); bare part of tail about 125 (4.90) wide, by 290 (11.40) long. Weight 40 to 60 pounds; female smaller.

Type locality, San Pedro River, near Monument 98 on the Arizona-Sonora boundary line.

Broad-tailed Beavers are found in the interior southwestern United States and northern Mexico from Sonora to Montana. Those found in eastern California along the Colorado River are of this subspecies.

In February and March, 1903, I saw signs of Beavers along the banks of the Colorado River a few miles below old Fort Yuma, but failed to get any in the traps which I set for them. They were few in number, probably only a pair, and seemed to choose a new place to come out on the bank each night. I found very few trees cut, these being mostly small willow saplings. The principal "signs" were at small, but dense, thickets of cane that grow here and there along the banks, and I saw some canes that had been cut. Beaver are known to live in suitable places all along the Colorado River, but they are trapped so persistently that they do not get a chance to become plentiful.

Castor canadensis pacificus RHOADS.
PACIFIC BEAVER.

Underfur of upper surface of body and head seal brown; overhair glossy reddish chestnut, almost concealing the underfur along the back; underfur of belly drab gray at roots and overhair broccoli brown; fore legs and feet dark wood brown; hind feet seal brown; ears black.

Length of type specimen (a female) 1145 mm. (45 inches); tail vertebræ 330 (14); hind foot 185 (7.25); bare part of tail 122 (4.80) wide by 295 (11.60) long.

Type locality, Lake Kichelos, Washington.

Pacific slope from Alaska to central California east to and including the Sierra Nevada and Cascade Mountains. I saw old Beaver dams and aspens cut by Beavers a few miles east of Goose Lake, Modoc County, but all the Beavers had been caught a few years previously. Dr. Cooper says that Beavers were formerly common in the San Joaquin and Sacramento Rivers, and I have reasons for believing that they are not exterminated there yet, though rare.

I can find no records of any "houses" having been seen in California and I have seen none west of Colorado. In places where dry banks that Beavers can burrow in occur the Beavers do not build houses. In fact all houses that I ever saw were placed in ponds made by damming streams so as to get still water to build in, and these localities were either too rocky to burrow in easily or suitable dry banks were not available. Dams are not often built in streams that do not freeze over, the principal use of the dam being to provide deep water to store logs and branches in for a food supply when the streams are frozen over and it is not practicable to cut wood and float it to where they wish to eat the bark and twigs. In most parts of Califorina the presence of Beavers is only made known by the stumps of the trees and saplings that they have cut.

The use of the tail as a trowel or barge is but another of these "fairy tales" that unfortunately creep into natural history accounts. The use of the tail in water is in diving and to some extent as a rudder. When on land it is used as a prop when the animal wishes to sit up and gnaw the bark from a stick held in the fore paws, or to cut down a tree. Swimming is done with hind feet, the fore feet being mostly held folded back under the breast. When swimming on the surface, if frightened or suspicious, it is not unusual for the Beaver to strike the surface of the water with the flat tail, making a sharp report, that heard near one on a still night is startling enough, as I know from experience. Beavers have been credited with great intelligence, but the facts do not indicate an uncommon account of reasoning power; many other rodents are nearly or quite as cunning.

Family **Muridæ.** Rats and Mice.

Skull much contracted interorbitally; anteorbital foramen large, wide in its upper part, narrow at bottom; zygomatic arch spreading, slender, the maxillar part prolonged backward and the malar correspondingly diminished; no premolars; molars rooted or rootless, tuberculate or with angular enamel folds on grinding surface; no external cheek pouches; internal cheek pouches sometimes present; clavicles present; tibia and fibula united in their lower parts; inner toe of front foot rudimentary.

This is a large family of nearly fifty genera and probably five hundred species divided among several subfamilies. The family is represented in all parts of the world, but each family preponderates in some particular zoo-geographical region.

Few members of this family are utilized by mankind as food. Taken as a whole it may be classed as noxious through their destroying considerable amounts of cultivated or indigenous crops or their stored products. Rats and Mice are more or less omnivorous. Perhaps their largest item of food is seeds, but scarcely anything edible comes amiss to some or another of the species.

Most of the species are nocturnal. The modes of life are varied; some are semiaquatic; a few are semiarboreal; most species are terrestrial and again others are more or less subterranean. The sexes are practically alike; the young are usually darker than the adults; distinct seasonal changes are few.

Dental formula, I, 1—1; C, 0—0; P, 0—0; M, 3—3×2=16.

Subfamily **Murinæ.**

Skull long and narrow; rostrum long; nasals projecting beyond premaxillaries; enlargement at root of lower incisor near base of condylar process greatest on the outer surface; tip of angular process below the plane of the summits of lower molars; notch between tip of angular process of lower jaw and condyle shallow; molars rooted, tuberculate, with tubercles in three series; palate extending further back than molars.

Genus **Mus** Linn. (Mouse.)

Incisors narrow, not grooved in front; tail long, nearly naked, the short sparse hairs not hiding the rings of scales covering it; ears rather large; pelage usually harsh.

Mus norvegicus Erxleben. (Of Norway.)
BROWN RAT.

Tail shorter than head and body; color above rusty brown thickly mixed with coarse black hairs; sides grayer; below ashy white; tail dusky, slightly paler below.

Length about 400 mm. (15.75) inches); tail vertebræ 190 (7.50); hind foot 42 (1.65).

Brown Rats were originally from central Asia, whence they spread to Europe. They were incidentally introduced into America in 1775. They have been known on the Pacific coast more than fifty years, coming ashore from shipping and gradually spreading through the country, but are yet unknown in many parts of the State distant from large towns. They inhabit towns preferably and are seldom seen far from buildings, in and under which they find shelter.

They are omnivorous and are great nuisances about barns, warehouses and dwellings. They are hardy, courageous and wary. They are more pugnacious than our native rats and soon drive the latter away from their neighborhood. They are sometimes called Norway Rats and Wharf Rats.

Mus rattus Linn. (Rats.)
BLACK RAT

Tail about as long as head and body; above sooty black; below plumbeous; feet brown; averaging smaller than *norvegicus*.

Introduced from Europe earlier than the Brown Rat, but overpowered by the latter and now rare. The habits of the two species are similar.

Mus musculus LINN. (Little Mouse.)
COMMON MOUSE.

Tail longer than head and body; above yellowish brown thickly mixed with black hairs; below ashy brown; feet brown; tail dusky, sometimes lighter below.

Length about 160 mm. (6.30 inches); tail vertebræ 82 (3.25); hind foot 18 (.70).

Introduced from Europe. Now found in most old settlements in the State. Principally frequent houses and other buildings, from which they drive the less objectionable native mice.

Subfamily **Cricetinæ.**

Skull short and moderately broad; rostrum rather short; nasals projecting beyond premaxillaries; enlargement at root of lower incisor near base of condylar process greatest on outer surface; tip of angular process of lower jaw below plane of summits of lower molars; notch between tip of angular process and condyle shallow; molars rooted, tuberculate, the tubercles in two series; palate ending opposite end of molar row.

Genus **Onychomys** BAIRD. (Claw—mouse.)

Upper incisors broad, causing a broadening of the rostrum at their roots; posterior molars above and below much smaller than the others; nasals long, wedge shaped posteriorly; coronoid process of lower jaw long, slender, curved backward; fore feet large with long claws; tail thick, blunt, short, about half as long as head and body.

Onychomys torridus ramona RHOADS. (Torrid; for "Ramona.")
SAN BERNARDINO GRASSHOPPER MOUSE.

A broad indefinite dorsal band from nose to tail dark brown; sometimes blackish; sides reddish bistre; below white, this color

including the feet, sides of face nearly to the level of the eyes
and nose; upper third of tail similar to the back, the remainder
white, usually including the tip; nasals long and pointed; a more
or less distinct supraorbital bead.   Immature; mouse gray above.

Length about 140 mm.   (5.50 inches); tail vertebræ 53
(2.10); hind foot 20 (.80); ear from crown 15 (.60).

Type locality, San Bernardino Valley, California.

San Bernardino Grasshopper Mice inhabit the valleys of
southwestern California and northwestern Lower California. They
are more frequently found in sandy land in valleys, but are nowhere common.   I have taken them along the seashore and in
the foothills, but not in the mountains.   They are more carnivorous than is usual with this family, the food consisting of
insects, such as grasshoppers, beetles and larvæ.   They attack
other mice and often devour parts of such mice as they find
caught in the collectors' traps.   They take grain bait, but meat
bait is preferred.   They have a musky odor.   They decay more
readily than common mice, probabl because of their carnivorous
diet.   The young are about four in number and are born in
March, April, May and June.   The mammæ are six in number,
one pair pectoral and two pairs inguinal.

Onychomys torridus perpallidus MEARNS.   (Very pale.)
YUMA GRASSHOPPER MOUSE.

Pelage long and soft; above vinaceous cinnamon, the hairs
tipped with black, sometimes producing a dark dorsal band; nose,
face nearly to eyes, feet and belly white; basal three fourths of
tail on the upper side mixed dusky and white; tip and underside
of tail white.

Length about 155 mm.   (6 inches); tail vertebræ 57 (2.25);
hind foot 22 (.87); ear from crown 16 (.63).

Type locality, Boundary Monument No. 204 (below Yuma,
Arizona).

The Yuma Grasshopper Mice seem to be local in distribution, and are common in a few places in Arizona, but rare on the California side of the Colorado River. Herbert Brown found them about Yuma in bottom lands thickly overgrown with weeds and cockle burs.

Onychomys torridus tularensis MERRIAM. (Of Tulare.)
TULARE GRASSHOPPER MOUSE.

Small; above pale drab gray barely tinged with buffy.

Length about 143 mm. (5.65 inches); tail vertebræ 50 (2); hind foot 21 (.83).

Type locality, Bakersfield, California.

Range, the Tulare Basin and vicinity; apparently not common.

Onychomys torridus longicaudus MERRIAM. (Long—tail.)
LONG-TAILED GRASSHOPPER MOUSE.

"Above cinnamon-fawn well mixed with black tipped hairs; ears small".

Length 145 mm. (5.70 inches); tail vertebræ 55 (2.15); hind foot 20 (.78); ear from crown 10 (.40); in dry skin.

Type locality, St. George, Utah.

An *Onychomys* occurs from Owen Valley and Death Valley eastward which I suppose to be *longicaudus,* but having no examples I may be mistaken in the species.

Genus **Peromyscus** GLOGER. (Pouch—little mouse.)

Upper incisors narrow; posterior molars somewhat smaller than the others; coronoid process of lower jaw small and low; tail tapering, shorter than head and body in some species, longer in others; pelage not harsh nor bristly.

Peromyscus texanus gambeli BAIRD. (For Dr. Wm. Gambel.)
GAMBEL MOUSE.

Tail shorter than head and body. Above variable in color

from light grayish wood brown to dark drab or hair brown, darkest along the back and top of the shoulders; feet and lower parts from nose to tail white; tail distinctly bicolor, the upper third brown or dusky, the remainder white. Occasionally a reddish or fawn colored individual is found; these are usually old animals. *Young;* mouse gray, scarcely lighter on the sides; belly grayish or ashy.

Length about 160 mm. (6.30 inches); tail vertebræ 74 (2.90); hind foot 20 (.80); ear from crown 17 (.67).

Type locality, Monterey, California.

The Gambel Mice are generally distributed from northern Lower California to Oregon, and from the western border of the Deserts west to the seacoast. They are found in the greatest variety of situations from the seacoast to timberline in the high mountains. They are perhaps less fond of brushy localities than several other species of the genus and frequent rocky localities more than they do.

The food consists of a great variety of seeds, leaves, twigs, bark, insects or flesh of any kind that may fall in their way. The young are born at all times of the year except in the coldest part of the winter; they are four to eight in number. The nests are warm masses of grass, sometimes lined with hair or feathers, and are placed in crevices among rocks, hollows in trees, or in burrows in the ground. The young are nearly hairless when born and are blind, the eyes not opening for several days.

This species frequents houses and other buildings in regions where the introduced house mouse has not become common. They are easily trapped in almost any kind of trap baited with grain, bread or fresh meat.

Peromyscus texanus deserticolus MEARNS. (Desert inhabiting.)
DESERT DEER MOUSE.

Pale; above yellowish drab, the sides tinged with ochraceous;

feet and lower parts white; tail bicolor, dusky or brownish above; remainder white.

Length about 180 mm. (7.10) inches); tail vertebræ 82 (3.25); hind foot 21 (.83); ear from crown 16 (.63).

Type locality, Mojave Desert near Hesperia, California.

Desert Mice are common in the arid regions of northeastern Lower California, southeastern California, southern Nevada, southwestern Utah and western Arizona, in all places where they can find food. Like most mammals of this arid region they are independent of water, though probably using it when it is to be had. Like other mice the food is varied.

The label of one of my skins from Salt Creek, Colorado Desert, taken March 29th, bears the notes "contained eight fœtuses". I have other fœtal notes in April (seven), June (five) and November (five). The habits of this subspecies do not differ materially from those of others of the species. Examples from parts of the San Bernardino and other mountains are very similar to *deserticolus* and perhaps should be referred to that form.

Peromyscus ~~texanus~~ *maniculatis* clementis MEARNS. (Of San Clemente Island.)
SAN CLEMENTE MOUSE.

"Above drab anteriorly, strongly tinged with burnt umber posteriorly; top of head drab gray; ears black with faint hoary edging; feet and under surface white; tail sharply bicolored".

"Length 177 mm. (7 inches); tail vertebræ 77 (3); hind foot 21 (.83); ear 17 (.67)".

Type locality, San Clemente Island, California.

I have seen a few *Peromyscus* from each of the following Islands, Coronado, San Clemente, San Nicolas, and Santa Barbara. These agreed in size but differed slightly in shade of color, the Santa Barbara skins being the darkest.

Peromyscus oreas rubidus Osgood. (A mountain nymph; red.)
MENDOCINO MOUSE.

"Upper parts brownish fawn with an evident median dorsal line; sides brownish fawn; ears lightly edged with whitish; under parts white; tail sharply bicolor."

Length about 193 mm. (7.62 inches); tail vertebræ 96 (3.80); hind foot 21.5 (.85).

Type locality, Mendocino City, Mendocino County, California.

"Coast region of northern California and southern Oregon, south at least to Cazadero, California."

Peromyscus boylii Baird. (For Dr. C. C. Boyle.)
BOYLE MOUSE.

Size medium; tail longer than the head and body; ears of moderate size; above varying from bistre mixed with blackish to mouse gray, the bistre specimens having the sides of head and sides of body strongly tinged with wood brown, the gray ones with very little reddish on the sides; tail bicolor, dusky above whitish below. *Immature;* slate gray above; pale ashy below.

Length about 195 mm. (7.70 inches); tail vertebræ 105 (4.15); hind foot 22 (.86); ear from crown 19 (.75).

Type locality, Middle Fork of American River, El Dorado County, California.

Boyle Mice are found in many parts of California, principally in the mountains, seldom occurring in the valleys. They are not often plentiful, and are occasionally found in houses and barns. They appear to be a brush loving species.

Peromyscus truei Shufeldt. (For F. W. True of the National Museum.)
BIG-EARED MOUSE.

Similar in colors to *boylii* and *californicus,* but averaging browner; body stout; tail distinctly bicolor; ears and hind feet long.

Length about 195 mm. (7.70 inches); tail vertebræ 105 (4.15); hind foot 23 (.90); ear from crown 23 (.90).

Type locality, Fort Wingate, New Mexico.

Big-eared Mice, including the subspecies, are widely distributed over the southwestern United States. In California they are found principally in the foothills and mountains of the coast region. They seem to be local in distribution and are probably most plentiful in the west central part of the State. Probably the Californian form will ultimately be known as subspecies *gilberti*. I have no material from the type locality and cannot be sure therefore that such is the real status of the Californian form.

I have most frequently found the Big-eared Mice in thickets of brush in open forest. The litters of young are small, usually but two or three in number.

Subgenus **Haplomylomys** Osgood. (Simple—molar—mouse.)

Skull with cranium relatively large; first and second upper molars with but two reentrant angles on the outer side, the small secondary tubercles being absent; lower molars correspondingly simple; tail longer than head and body, thinly haired.

Peromyscus californicus Gambel. (Of California.)
CALIFORNIA MOUSE.

· Size very large; ears very large; tail long, short haired, distinctly bicolor in adults; soles naked; above yellowish brown thickly mixed with black, especially on the back and hips which are often nearly black; sides tinged with ochraceous passing to ochraceous buff on the lower part of the sides and there strongly contrasting with the grayish white lower parts; breast more or less tinged with ochraceous, often forming a spot; feet white;

tail blackish above, dull white below, sometimes tipped with white. *Young;* plumbeous above, blackish at a later stage, with but little ochraceous tinge on the sides; below ashy or grayish white; tail scarcely lighter beneath.

Length about 250 mm. (9.85 inches); tail vertebræ 140 (5.50); hind foot 26 (1.03); ear from crown 23 (.90).

Type locality. Monterey, California.

The California Mouse is found in the chemisal and in underbrush in open forests, in the valleys, foothills and lower mountains of the coast region of California from some distance north of San Francisco south to about Santa Barbara where it blends into the next subspecies. It appears to reach the lower part of the Sierra Nevada in small numbers.

For some time after its discovery naturalists supposed that this species lived in the nests of the Brush Rats and were in some manner parasitic on them, but it is now known to occur in brush in general and to have habits similar to those of other wood mice.

Peromyscus californicus insignis RHOADS. (Distinguished by a mark.)
CHEMISAL MOUSE.

Very similar to *californicus;* slightly smaller in average; lighter colored, back with less black, sides less ochraceous.

Length about 233 mm. (9.85 inches); tail vertebræ 130 (5.10); hind foot 25 (1); ear from crown 22 (.86).

Type locality, Dulzura, San Diego County, California.

The Chemisal Mouse occurs in northwestern Lower California and in southern California from the seacoast to the lower edge of the pines. They do not frequent open valleys, but are more or less common in the chemisal and in the brush among the oaks. They frequently inhabit knotholes and hollows in leaning trees, being fair climbers. They are fond of running on logs. I do not find them more common about Brush Rat nests than elsewhere. They appear to breed at all times of year. The litters are small, oftenest consisting of three young.

Peromyscus eremicus BAIRD. (Hermit.)
HERMIT MOUSE.

Pale colored; tail long, very slender, scant haired; soles naked; ears large; above broccoli brown, grayer on the head, mixed with black hairs on the back; sides ochraceous buff, strongest on the lower part of the sides; belly white, distinctly outlined against the buff sides; feet white; tail dusky above, pale gray below, but not distinctly bicolored. *Young:* darker, with little or no buff on the sides.

Length about 195 mm. (7.70 inches); tail vertebræ 107 (4.20); hind foot 21 (.83); ear from crown 17 (.67).

Type locality, old Fort Yuma, California.

Hermit Mice are generally distributed through the eastern parts of the Colorado and Mojave Deserts, the valleys and deserts of western Arizona, southern Utah, southern Nevada, Sonora and northeastern Lower California. They are perhaps most common in rocky ground in the hills and barren mountains of this region, occurring up to 4,000 feet altitude; but they are also occasionally common miles out on the plains in the rare patches of grass and weeds and in the vegetation about springs or the sinks of the infrequent springs. Their food is mostly seeds, but beetles and other insects are also eaten. The young number three or four in a litter.

Peromyscus eremicus stephensi MEARNS (For F. Stephens.)
PALM DESERT MOUSE.

Very similar to *eremicus;* averaging smaller with proportionally longer tail; paler; belly white.

Type locality, canyon below Mountain Spring near the Mexican boundary, San Diego County, California.

Rather common in the foothills along the western border of the Colorado Desert.

Peromyscus eremicus herroni RHOADS. (For R. B. Herron.)
HERRON MOUSE.

Similar to *eremicus;* darker; sides less buffy, the color of the back shading further down the sides; belly grayish white.

Type locality, south side of San Bernardino Valley, California.

Herron Mice are intermediate between the Hermit Mice and Dulzura Mice in color and in habitat. They live in the drier warm interior valleys, and their borders in the southern part of California, from northern San Diego County northward. They are common in few places.

Peromyscus eremicus fraterculus MILLER. (Little brother.)
DULZURA MOUSE.

Darker than *eremicus* or *herroni;* above dark grayish wood brown or yellowish bistre rather thickly intermixed with black, shading on the sides to brownish ochraceous buff; belly cream buff in typical specimens but often pale grayish buff or grayish white; a buff pectoral spot is frequently present.

Type locality, Dulzura, San Diego County, California.

Rather common in brush along the coast and mesas and western slopes of the coast mountains from northwestern Lower California northwest to Ventura County, California. The northernmost specimens intergrade with Herron Mice and the eastern ones with Palm Desert Mice.

Genus Sigmodon SAY and ORD. (Sigma—tooth.)

Upper incisors broad; rostrum broad and short; zygomatic arches very wide posteriorly; border of orbits beaded; coronoid process of lower jaw of moderate size; a process of the maxillary

projecting in front of the anteorbital foramen nearly cutting it in
two; tail slender, scaly, thinly haired, shorter than head and
body; pelage long, coarse, hispid; form stout.

Sigmodon hispidus eremicus MEARNS. (Bristly; hermit)
WESTERN COTTON RAT.

Above grayish buff coarsely grizzled with black, paler on
the sides; below dull white, the plumbeous bases of the hairs
showing through the white tips; feet grayish white; tail blackish
above, grayish below.

Length about 280 mm. (11 inches); tail vertebræ 130
(5.10); hind foot 34 (1.35); ear from crown 17 (.67).

Type locality, northwestern Sonora, Mexico, 30 miles south
of boundary monument 204, near the Colorado River.

Western Cotton Rats are found in the bottom lands of the
Colorado River from its mouth north to near Ehrenberg, Arizona
or further. But little is known about their abundance, but they are
probably common in places, and are likely to prove troublesome
as settlements increase and food and cover become more plentiful. They seem to like thick cover such as cane patches and thick
weeds, and are likely to invade grain and alfalfa fields.

They are prolific, as I caught females opposite Ehrenberg
in August containing six fœtuses each. None of the females
that I caught near Yuma in March contained any. Their habits
seem to be similar to those of meadow mice in some respects. I
found sorghum stalks cut in coarse pieces six to ten inches long.

Genus **Reithrodontomys** GIGLIOLI. (Channel—tooth—
mouse.)

Upper incisors deeply grooved in front, appearing collectively as if there were four instead of two; lower incisors small,
normal; front upper molar with four roots, one being very small;
coronoid process of lower jaw small, oblique; angular process in-
flected at lower edge; anteorbital foramen wide and rounded

above, contracted to a slit below; tail usually longer than head and body, slender, moderately haired.

Reithrodontomys longicaudus BAIRD. (Long—tail.)
LONG-TAILED HARVEST MOUSE.

Adult; above reddish bistre thickly mixed with black hairs, these usually forming a broad blackish dorsal band; sides with fewer black hairs and more or less tinged with cinnamon; below grayish white, sometimes tinged with buff; tail indistinctly bicolor, dusky above, whitish below. *Immature;* mouse gray above, pale plumbeous below.

(I consider *pallidus* Rhoads not separable from *longicaudus.* If *pallidus* is recognized as a subspecies at least two more subspecies must be named, but I do not think these slight local differences sufficiently tangible to the worth recognizing).

Length about 143 mm. (5.63 inches); tail vertebræ 76 (3); hind foot 17 (.67); ear from crown 13 (.51).

Type locality, Petaluma, California.

Long-tailed Harvest Mice are found from Lake and Tehama Counties south into northern Lower California; and from the seacoast east into the Sierra Nevada and San Bernardino Mountains. They are found in grassy localities. The thicker and older the grass the more abundant the Mice are likely to be. In a few localities they are quite common, but they may be wanting over large areas. They do not appear to go high in the mountains, seldom as high as 4,000 feet altitude.

The food seems to be entirely vegetable, mostly the seeds, leaves and stems of various plants. The mammæ are six in number, two pairs inguineal and one pair pectoral. The young are two to four and are born at all times of the year, probably two or three litters annually.

Reithrodontomys megalotis deserti ALLEN. (Large ear; of the desert.)
DESERT HARVEST MOUSE.

Similar to *longicaudus;* grayer, with fewer black hairs mixed through the pelage of the back; ears broader and averaging higher.

Length about 140 mm. (5.50 inches); tail vertebræ 73 (2.87); hind foot 17 (.67).

Type locality, Oasis Valley, southwestern Nevada.

Desert Harvest Mice occur in patches of grass and weeds around springs and the infrequent small streams of southern Nevada and the adjoining part of California, west to the foot of the Sierra Nevada, and south to northeastern Lower California. Their habits are similar to those of the Long-tailed Harvest Mice, but they reach a somewhat higher altitude.

Reithrodontomys klamathensis MERRIAM. (Of Klamath Valley.)
KLAMATH HARVEST MOUSE.

"Upper parts pale grayish brown, washed with buffy on sides; under parts white; tail bicolor, dusky above, whitish below; ears and hind feet large.'

Length 144 mm. (5.70 inches); tail vertebræ 66 (2.60); hind foot 18.5 (.73).

Type locality, Shasta Valley, California.

Said to be rather common from the base of Mount Shasta north and northeast. I have not seen this species.

Subfamily **Neotominæ.**

Skull long and narrow; rostrum long; nasals projecting beyond premaxillaries; enlargement at foot of lower incisor near base of condylar process greatest on outer surface; tip of angular process of lower jaw below plane of summits of molars; notch between tip of angular process and condyle very shallow; molars rooted or semirooted, prismatic; palate ending about the middle of last molars.

Genus **Neotoma** Say and Ord. (New—to cut.)

Upper molars with three roots, lower with two; last molar smallest; coronoid process slender, usually higher than condyle; anteorbital foramen wide above, much contracted below, the maxillar plate bounding its posterior side not spurred; frontal not distinctly beaded at border of orbit; audital bullæ small; eyes prominent; ears large, rounded; thinly haired; whiskers very long; size large.

Subgenus **Neotoma** Gray.

Skull strong, rugged; rostrum elongated; tail broad, squirrel-like; hind feet large.

Neotoma cinerea Ord. (Ash gray.)
ASH-COLORED RAT.

Above mixed yellowish brown and black, sides with more buff and less black; below white, the hairs ashy at base except on the breast; feet white; ankles dusky; tail rather darker on the upper side than the back, the yellow tints lacking, below white except near the base where it is brown. The hairs of the tail are from a quarter of an inch to a full inch in length, varying in length with age, season and individual. *Young;* above slate gray thickly mixed with black hairs; below ashy white; tail ashy or slate gray above, white below, hairs short but longer than in young of the subgenus *Neotoma.*

Length about 380 mm. (15 inches); tail vertebræ 180 (7); hind foot 45 (1.75).

Type locality, Great Falls, Montana.

Northern Rocky Mountains and west and southwest to the Cascade Mountains and Sierra Nevada. Common in Modoc and

Lassen Counties and occasional in the Sierras south to Mount Whitney. Occasional on Mount Shasta and probably occur in small numbers in the mountains west to the Pacific.

Ash-colored Rats appear to live mostly among rocks, often in lava cliffs. The nests are not nearly as large as those of *fuscipes* and some others. In trapping for them I succeeded best with meat baits. The food is varied but is mostly vegetable, including juniper berries and twigs. In winter they invade barns and houses, and carry off anything eatable and many uneatable things that take their fancy. A peculiarity not frequent in this genus is the strong musky odor, which remains with skins in the cabinet many years. Judging from the scanty material at hand the young are born from the first of May to the end of July.

Subgenus **Neotoma**.

Skull comparatively smooth and thin; rostrum of moderate length; tail scant haired, rat-like; hind feet of moderate size.

Neotoma fuscipes BAIRD. (Dusky—foot.)
DUSKY-FOOTED BRUSH RAT.

Large; tail long; ears large; above bistre or sepia darkened by black tips of the hairs, base of hairs slaty; sides varying from grayish tawny olive to grayish brown, shading into the color of the back, distinctly outlined against the grayish white or buffy white belly and throat, the hairs of the lower parts plumbeous at base except on throat, breast and anal region; fore feet and toes of hind feet white, the upper surface of the hind feet dusky or spotted with dusky; ankles blackish; tail blackish scarcely lighter beneath; hairs of tail short but hiding the skin. *Young;* gray with very little tawny or reddish tinge.

Length about 407 mm. (16 inches); tail vertebræ 205 (8.10); hind foot 40 (1.60); ear from crown 35 (1.40).

Type locality, Petaluma and Santa Clara, California.

Pacific coast region of central California from Monterey County north to Lake County. Dusky-footed Brush-Rats inhabit the chemisal and the underbrush in open forests and groves, rarely being found in thick forests. This form does not appear to occur high in the mountains, seldom up to 3,000 feet altitude. The food is principally vegetable but it is quite varied. They have the usual generic propensity for carrying off small articles.

The breeding season is March to June, perhaps later. The number of young in a litter is two to four. The home is usually in a "nest" or "house" of sticks, twigs, bones, or anything portable; these piles of rubbish being two to four feet high, roughly cone shaped, and are usually placed in a thicket of brush, sometimes against a tree.

Occasionally the Brush-Rats take up their residence in barns or other buildings where they do the most harm by carrying off small articles, stored vegetables, dried fruit, grain or anything they can carry off, even if utterly useless to them except to swell their rubbish pile. They seldom gnaw anything, however. They leave the premises immediately on the arrival of the introduced species of rat, which is a greater nuisance.

Neotoma fuscipes monochroura RHOADS. (One-color—tail.)

NORTHERN DUSKY-FOOTED BRUSH-RAT.

Similar to *fuscipes;* darker above; hairs of belly white to roots; skull flatter; molar tooth row shorter.

Type locality, Grant Pass, Josephine County, Oregon.

Pacific coast region from Mendocino County, California north to mouth of the Columbia River, east to base of Mount Shasta.

Neotoma fuscipes marcotis THOMAS. (Large—ear.)

SOUTHERN BRUSH RAT.

Similar to *fuscipes;* grayer, with less fulvous on the sides;

tail bicolor, blackish above, grayish below; upper surface of hind feet more or less clouded with dusky; averaging smaller; tail shorter proportionally; palate usually shorter than incisive foramina.

Length about 380 mm. (15 inches); tail vertebræ 190 (7.50); hind foot 37 (1.45); ear from crown 29 (1.15).

Type locality, San Diego, Califarnia.

Southwestern California and northwestern Lower California, from the seacoast up to 7,000 feet altitude in the mountains. The Southern Brush-Rat is found in chemisal and other brush. The nests are large and may be seen frequently in suitable places. Occasionally smaller nests are placed in trees which lean. These tree nests are probably used in warm weather, and are commonly near other nests on the ground. The habits in general are the same as those of the species elsewhere.

Neotoma fuscipes simplex. TRUE. (Simple.)
XANTUS BRUSH RAT.

Similar to *macrotis;* smaller and grayer; hairs of lower parts white to roots; hind feet white; tail bicolor.

Type locality, old Fort Tejon, California.

Foothills and mountains bordering the southern part of the San Joaquin Valley and the extreme western part of the Mojave Desert.

Neotoma fuscipes streatori MERRIAM. (For C. P. Streator.)
STREATOR BRUSH RAT.

Similar to *fuscipes* in size and color; ankles darker; hind foot from ankle pure white; tail bicolor, blackish above, whitish below; skull somewhat different from that of *fuscipes;* length of palate less than that of incisive foramina, which reach back somewhat beyond the front of the first molars; zygomatic arches less spreading posteriorly.

Length about 380 mm. (15 inches); tail vertebræ 183 (7.20); hind foot 37 (1.45.)

Type locality, Carbondale, Amador County, California.

Western slope of the Sierra Nevada and northeastern California.

Neotoma fuscipes dispar MERRIAM.. (Below par, degraded.)
PALE BRUSH RAT.

Entire upper parts ochraceous buff, palest on the head; back moderately lined with black tipped hairs; feet and under parts white; the white of the belly encroached upon the buffy ochraceous of the sides; tail bicolor, above brownish gray, below soiled white. The skull is similar to that of *streatori*.

Type locality, Lone Pine, Inyo County, California.

Eastern foothills of the Sierra Nevada from Owen Valley southward to the Mojave Desert. This subspecies seems to be rare. They are similar to *desertorum* in color but may be known by the dark ankles and long tail.

Neotoma desertorum MERRIAM. (Of the desert.)
DESERT BRUSH RAT.

Above brownish buff darkened by a mixture of black hairs, grayer on the head, clearer buff on the sides, which are usually strongly contrasted against the white lower parts; feet white; tail bicolor, dusky above whitish below. There is some variation in color, examples from some localities being paler or more buffy above, and the lower parts may be tinged with buff, especially across the breast. The pelage is very soft and the tail is considerably shorter than the length of head and body.

Length about 290 mm. (11.40 inches); tail vertebræ 135 (5.30); hind foot 30 (1.20).

Type locality, Furnace Creek, Death Valley, California.

Mojave Desert, southern Nevada and southwestern Utah.

Common in many parts of this region, more especially in rocky localities. The food is almost anything eatable, but from the nature of the region they live in, this is mostly limited to the leaves, twigs, bark and seeds of desert plants, including cactuses. The nests are commonly placed in crevices among rocks, or under cactuses or yuccas; these very frequently contain thorny twigs and joints of cactuses, and are sometimes built exclusively of such formidable materials, perhaps for protection against coyotes.

The young are three to five in number. My notes on fœtuses observed include only March and April as breeding months, but the season is probably longer than these indicate. The Desert Brush-Rats have the usual thieving habits of the genus, as many prospectors can testify, bright objects being especially attractive. I find it nearly useless to put out any "cyclone" traps near their nests, the tin bottoms proving too attractive.

The Desert Brush-Rats were formerly a considerable item of food for the Indians, but they use them less now, partly because other food has become available, but principally because of the ridicule of the whites. The flesh is sweet, white and nutritious, and there is no good reason why it should not be as palatable as that of a squirrel. The *Neotomas* are very different from real rats.

Neotoma desertorum sola MERRIAM. (Alone.)
KERN BRUSH RAT.

Similar to *desertorum* but larger.

Length about 325 mm. (12.80 inches); tai vertebræ 150 (5.90); hind foot 34 (1.35).

Type locality, San Emigdio, Kern County, California.

Distribution, head of San Joaquin Valley, California.

Neotoma intermedia RHOADS. (In the middle.) .
INTERMEDIATE BRUSH RAT.

Similar to *desertorum* but darker with less buff on the sides; body scarcely larger but tail longer, nearly or quite as long as head and body; pelage soft but less so than in *desertorum;* above light buffy brown; sides lighter; below grayish white or buffy white; feet white; tail blackish above, whitish below; skull considerably larger than that of *desertorum,* heavier and more angular; interorbital constriction wider proportionally; incisive foramina terminating slightly posterior to plane of anterior edge of first molars. *Immature;* darker, with but little buff tinge.

Length about 305 mm. (12 inches); tail vertebræ 152 (6); hind foot 32 (1.25); ear from crown 28 (1.10).

Type locality, Dulzura, San Diego County, California.

Valleys and slopes of the coast region of southern California, north nearly to Monterey. Apparently not found much above 3,000 feet altitude. They prefer rocky localities and usually build their nests among rocks.

Neotoma intermedia gilva RHOADS. (Yellowish.)
YELLOW BRUSH RAT.

Very similar to *desertorum* in color but with the long tail and large skull of *intermedia.* Averaging a little smaller than *intermedia.*

Type locality, the San Gorgonio Pass, California.

Distribution, San Gorgonio Pass and the Colorado Desert.

The following notes on a mother and young are extracts from a letter to me from Mr. A. H. Alverson of San Bernardino. The locality given is the Desert end of the San Gorgonio Pass. "She was taken within a mile of Whitewater, in the low foothills. The nest was under a bunch of *Cereus engelmani,* but she was out and about two feet away from the entrance, which led to her discovery—the cause of her being out at that time of day I do not know, it being about 10 A. M. When she returned I noticed that she had young attached to her mammæ. I soon had the plant overturned and digging about a foot deep came upon her. One of the young—there were three—became detached and set up a lively squeaking. It soon got a small stick in its mouth and held on with considerable strength, but being placed with the mother soon found its proper hold, which they all seemed to maintain until the eyes were open or nearly so; then I noticed that when the mother desired to move to another part of the cage—she is very neat—she would turn round and round and seem to twist them loose in a pile, where they would lie quietly until they felt her return, then they would at once attach to the teats, which as you know are placed very far back.

"I found them about May 10th, and they may have been about a week old. About a week ago (about May 20th), their eyes began to open, and now they are wide open. They eat with the mother, who takes almost anything from roast beef or bacon, to seeds, fruit, or bread, and is very fond of milk. Water was first given her, which she lapped like a cat, long and often. I have them in a thin wooden box, with a glass front, for observation. She does not seem inclined to gnaw, is quiet and not afraid, comes to the glass when opened and takes food from my hand, does not try to dart out, nor bite. Sometimes, however, when without sufficient food, she becomes uneasy and gnaws at the wooden box for a short time, but when food is placed in the box she desists.

"The young now have fully opened eyes, eat everything the mother does, are very playful, running about most of the time, but when too venturesome the mother takes them in her mouth and lifts them bodily back to the nest in the corner, which consists of well shredded cotton cloth—done by herself. · Sometimes she lifts them by the neck, but mostly by the middle of the side. After playing and eating the mother and young make their toilet, the mother doing mostly for all, but the young try to learn; then the young attach to the mammæ and all sleep.

"They seem to be quite nocturnal, decidedly more active at night, and are out in the day only to eat a little. I should think the young are now about four weeks old. They are nearly half the size of the mother and are growing rapidly. The mother and young appear to be the same color. They feel a lower temperature, are then less lively and the hair is slightly raised, especially on the head."

Neotoma abigula venusta TRUE. (White—throat; beautiful.)

MESQUIT BRUSH RAT.

Above mixed dusky and ochraceous buff; darkest on the crown and back, lighter and more buffy on the sides; below white; feet white; tail bicolor, blackish above, dull white below; skull strong and angular; rostrum short, wide and deep, depressed; nasals wide and broadened anteriorly, narrowed to a wedge shape posteriorly; frontal shortened posteriorly and parietals correspondingly lengthened; incisive foramina short. *Young;* paler gray than usual in this genus.

Length about 370 mm. (14.50 inches); tail vertebræ 175 (6.90); hind foot 35 (1.40); ear from crown 30 (1.18).

Type locality, Carrizo Creek, California. (In foothills bordering the Colorado Desert.)

The Mesquit Brush-Rats are most common in shrubby masses of mesquit scattered through the Colorado Desert and in the Colorado Valley. They also occur some distance up the gulches and canyons of the adjoining foothills. Their principal food is the mesquit "beans" and twigs. They are less given to nest bulding than most Brush-Rats, living more in burrows under mesquit trees. The breeding season is similar to that of the genus in general.

Subfamily **Microtinæ.**

Skull short and broad; rostrum short; nasals short, not projecting beyond premaxillaries; enlargement at root of lower incisor near base of condylar process of jaw greatest on inner surface; angular process bent back and up until its tip reaches above the plane of the summits of lower molars; notch between tip of angular process and condyle deep; molars prismatic, usually root-less; size usually small or medium, large in one genus.

Genus **Phenacomys** MERRIAM. (Cheat—mouse.)

Skull strong and angular; molars of young animals rootless, those of adults rooted, strong, with sharp outer angles; cusps of lower molar largest on tongue side of teeth; basal part of lower incisor passing beneath roots of lower molars; feet normal; tail round, one third to one half the length of head and body; size small.

Phenacomys orophilus MERRIAM. (Mountain—loving.)
MOUNTAIN LEMMING MOUSE.

Above grayish brown, tinged with yellow in summer, thickly sprinkled with black hairs; belly dirty white; feet whitish; tail bicolor, mixed brown and white above, whitish below.

Length about 145 mm. (5.70 inches); tail vertebræ 35 (1.38); hind foot 18 (.70).

Type locality, Salmon River Mountains, Idaho.

Higher parts of the mountains of British Columbia and western United States south to Mount Shasta, where Walter K. Fisher caught three "in the heather meadows along the upper part of Squaw Creek."

Phenacomys albipes MERRIAM. (White—foot.)
REDWOOD LEMMING MOUSE.

Above grizzled bistre with brownish wash on head, shoulders and sides; sides of nose dark grayish plumbeous with buffy wash; feet white; ankles dusky; tail bicolor, dusky above, whitish below.

Length of type 168 mm. (6.60 inches); tail vertebræ 62 (2.45); hind foot 19 (.75).

Type locality, redwoods near Arcata, California.

Genus **Evotomys** Coues. (Good—ear—mouse.)

Skull thin and smooth; molars of young animals rootless, those of adults rooted, rather weak, with rounded outer angles; inner cusps of lower molars about equal to outer; basal part of lower incisor passing on tongue side of roots of first and second molars and on outer side of third; feet normal; tail round, one third to one-half as long as head and body; size small.

Evotomys californicus Merriam.
CALIFORNIA RED-BACKED MOUSE.

A broad indistinct band from eyes to rump sepia mixed with black and gray; sides grizzled grayish brown, shading into the whitish under parts, which are tinged with buffy and darkened by the plumbeous under fur showing through; tail bicolor, dusky above, light brown or whitish below; feet dull white.

Length about 150 mm. (6 inches); tail vertebræ 46 (1.80); hind foot 19 (.75); ear from crown 8 (.30). Oregon examples appear to be larger.

Type locality, Eureka, California.

This species inhabits the coast region of northern California and western Oregon. But few specimens have been seen yet and their habits are not very well known. I have taken several individuals in redwood forests in Mendocino County, and they probably occur further south. Mine were trapped on dry hillsides in thick forest, in traps set alongside old logs or at the roots of trees.

The habits of the California Red-backed Mice are probably like those of the rest of the genus, which live in cool moist forests and brush lands, and delight in deep shade and the cover of logs, leaves and tangled weeds. Nests are built under logs, in underground burrows, or under cover of old leaves. Though mainly nocturnal some species are sometimes seen in the daytime. All sorts of seeds and green vegetation are eaten, and probably some worms and insects.

Evotomys obscurus Merriam. (Dusky.)
DUSKY RED-BACKED MOUSE.

Above olive gray with an ill-defined dorsal area of cinnamon rufous obscured by black hairs; lower part of sides and ' face clear gray; tail bicolor, dusky above, whitish beneath.

Length about 148 mm. (3.80 inches); tail vertebræ 46 (1.80); hind foot 17 (.67).

Type locality, Prospect, Upper Rogue River Valley, Oregon.

West slope of the northern Sierra Nevada and southern Cascade Mountains.

Evotomys mazama Merriam.
CRATER LAKE RED-BACKED MOUSE.

Dorsal stripe from in front of ears to base of tail cinnamon rufous or hazel, shading gradually into buffy gray on sides and face; belly washed with buffy white; tail sharply bicolor, blackish above, whitish below.

Length about 157 mm. (6.20 inches); tail vertebræ 51 (2); hind foot 19 (.75).

Type locality, Crater Lake, Oregon.

Higher parts of the Cascade Mountains in Oregon and Mount Shasta, California, between 5,000 and 8,000 feet altitude.

Genus Microtus Schrank. (Small—ear—mouse.)

Skull strong and angular; molars rootless through life, strong, with sharp outer angles; outer and inner cusps of lower molars of about the same size; basal part of lower incisor passing on tongue side of the bases of first and second molars and on outer side of third; feet normal; tail round, usually less than half as long as head and body; form stout; size medium or small.

Subgenus **Microtus**.

Pelage long and rather coarse; soles with six tubercles; pattern of enamel folds of third lower molar without closed triangles; third upper molar with three closed triangles and seven or eight salient angles.

Microtus montanus PEALE.   (Of the mountain.)
PEALE MEADOW MOUSE.

Nasals small, short, not projecting as far forward as the premaxillaries do; incisive foramina constricted posteriorly; pelage soft; above sepia mixed with black; below slate gray washed with white; tail scarcely one third as long as head and body, blackish above, lighter below; feet dull brown.

Length about 165 mm.   (6.50 inches); tail vertebræ 50 (2); hind foot 21 (.82); ear from crown 10 (.40).

Type locality, Sacramento River near Mount Shasta.

Peale Meadow-Mice inhabit meadows and marshes of the foothills and lower mountain sides of northeastern California, eastern Oregon, northern Nevada and Utah.   They do not seem to be common.

Microtus dutcheri BAILEY.   (For B. H. Dutcher.)
DUTCHER MEADOW MOUSE.

Similar to *montanus* in color; lips and usually tip of nose white; tail short; ears small, nearly concealed; nasals small and short; above sepia mixed with brown and black; below buffy brown (adult) or grayish (immature);   tail bicolor, blackish above, whitish below.

Length about 163 mm.   (6.40 inches); tail vertebræ 37 (1.45); hind foot 21 (.82).

Type locality, Big Cottonwood Meadows, 10,000 alt., near Mount Whitney, California.

Dutcher Meadow-Mice inhabit the wet valleys of the Sierra Nevada, from the head of Owen River southward, between 7,000 and 11,000 feet altitude. They are common in many of these valleys.

Microtus californicus PEALE. (Of California.)
CALIFORNIA MEADOW MOUSE.

Winter pelage, long and coarse; above wood brown or bistre darkened by intermixture of long black hairs on the back, basal two thirds of the pelage slaty black; sides grayer; below tipped with white the plumbeous under fur showing through; tail dark brown above, grayish below; feet light brown. *Summer pelage;* grayer; tail less distinctly bicolor.

 ̄Length about 170 mm. (6.70 inches); tail vertebræ 54 (2.10); hind foot 22.5 (.88); ear from crown 14 (.55).

Type locality, San Francisco Bay, California.

California Meadow-Mice occur from northern Lower California through southern and central California, west of the Colorado and Mojave Deserts, north along the coast to southwestern Oregon, and east into the western foothills of the Sierra Nevada. They are found in grassy localities, both dry and wet.

The food is stems and leaves of grasses and other plants, their roots and seeds, the bark of shrubs and trees when other food is not available, and probably some insects. They are sometimes destructive to grass and grain crops, but they are rarely as abundant in California as they are in colder climates. Many are caught by hawks, owls, skunks and other carnivorous animals. They are abroad more or less during the day, and the marsh hawk is perhaps their principal diurnal foe, while the barn owl destroys many of them in the night. California Meadow-Mice, like most of their genus, are in the habit of following regular paths. These runways ear easily found in thick grass by parting it and if these are numerous, the mice are abundant. A close inspection will show the stumps of grass and often little bunches of grass cut in short lengths can be found. A "cyclone" trap set in the runway so that the mouse will pass through it, or a small steel trap bedded so that the pan is level with the runway will usually prove successful. Where they are very abundant a narrow trench dug across the runway, a foot or so deep with straight sides, and visited night and morning will help thin them out. They are excellent swimmers so it is not easy to drown them. Their natural enemies are the most effectual means of keeping them in check. Protect such.

Three to eight young form a litter and several litters are born annually. I have taken these Meadow-Mice containing young nearly every month in the year. The young are born blind and almost hairless. The nests are placed under logs, stumps, in burrows, and sometimes in thick grass on the surface.

One clear September morning I was camped by the side of a brook in the mountains of San Diego County. The little stream in some winter flood had cut a channel in the alluvial soil five or six feet deep with nearly perpendicular banks and a dozen feet wide. For a short distance below camp the bottom of the channel was moist and overgrown with watercress and a few round tulles, through which the little stream meandered. In this vegetation some Meadow-Mice were feeding. I laid on the edge of the bank and watched them half an hour with the field glass, through which they appeared nearly within reach of my hand.

First some dry leaves were moved on a little slope at the bottom of the opposite bank and the head and back of No. 1 appeared. It was feeding on some small plants, but did not come out openly. Presently No. 2 ran out of the tulles on the watercress and began eating it. It moved in a nervous, jerky way, but did not appear shy. Soon No. 3 came but it was shyer and did not stay long, biting off a small tulle about a foot and a half long and dragging it into a thicker patch of tulles. It ran quickly as if accustomed to pulling such loads.

I did not see any of them sit up to eat, as many small mammals do, nor did they use their fore feet to hold their food, using only the mouth, apparently turning the leaves about with their tongue. They did not take the larger sprays of watercress.

They did not appear to chew the leaves much, but munched them down rapidly.

Their ears appeared rather prominent, considering the length of the surrounding pelage. The eyes were very prominent, like black beads, and had a staring expression. By nine o'clock they disappeared.

Microtus californicus vallicola BAILEY. (Of the valley.)
VALLEY MEADOW MOUSE.

Very similar to *californicus;* averaging larger and grayer.

Type locality, Lone Pine, Inyo County, California.

Marshy and grassy places in Owen Valley and the Mojave Desert west of Death Valley.

Microtus californicus constrictus BAILEY.
(Drawn together.)
COAST MEADOW MOUSE.

Averaging smaller and grayer than *californicus;* skull narrower; audital bullæ narrower; above buffy gray; below whitish; tail scarcely bicolor, dull grayish.

Type locality, Mendocino County, California.

Common on grassy hillsides and in pastures in the region along the coast near Cape Mendocino.

Microtus edax LE CONTE. (Voracious.)
TULLE MEADOW MOUSE.

Skull long, angular, heavily ridged; pelage blacker than in *californicus;* sides more grayish; feet large and stout.

Length about 215 mm. (8.45 inches;) tail vertebræ 70 2.75); hind foot 25 (1).

Type locality, near San Francisco, California.

Tulle swamps of the Sacramento and San Joaquin Valleys, California.

Microtus scirpensis BAILEY.   (Of the tulles, *Scirpus.*)
DESERT MEADOW MOUSE.

Similar to *edax* in size, proportions and skull; pelage grayer.

Known only, from a little tulle patch at a warm spring, near the Amargosa River, Inyo County, California, below the Nevada-California boundary.

Microtus mordax MERRIAM.   (Biting.)
CANTANKEROUS MEADOW MOUSE.

Pelage very coarse; color light; above pale bistre grizzled with gray and black; below whitish, the plumbeous underfur showing through; tail indistinctly bicolor, brownish above, below light gray.

Length about 180 mm.   (7.10 inches); tail vertebræ 64 (2.50); hind foot 22 (.87) ; ear from crown 14 (.55).

Type locality, Sawtooth Lake, Idaho.

From the Rocky Mountains through the ranges of the Great Basin to the Sierra Nevada, Mount Shasta and Trinity Mountains, south to the San Bernardino and San Jacinto Mountains. In California they are found only in high mountains, from 5,000 feet alt., in the northern part of the State and 7,000 feet in the southern part, up nearly to timber line.

Microtus angusticeps BAILEY.   (Short—head.)
BAILEY MEADOW MOUSE.

Above dark bistre mixed with black, darkest on face; below washed with creamy white; feet plumbeous gray; tail distinctly bicolor, blackish above, soiled white below; pelage coarse; skull small, narrow; audital bullæ small; molars small, with narrow, sharp angles.

,Length about 170 mm.   (6.70 inches) ; tail vertebræ 55 (2.15); hind foot 22 (.87).

Type locality, Crescent City, California.

Bailey Meadow-Mice occur in the damp pastures in the

Sitka spruce belt along the coast of northwestern California and southwestern Oregon.

Subgenus **Lagurus**.

Pelage long and rather coarse; soles with five tubercles; pattern of third lower molar with two or three closed triangles; third upper molar with two or three closed triangles and five or six salient angles; palate flattened; audital bullæ large and projecting backward.

Microtus curtatus COPE. (Shortened.)
SHORT-TAILED MEADOW MOUSE.

Above pale buffy gray; soiled white below; tail pale gray, slightly darker above; very short; skull wide and flat, with short rostrum; audital bullæ inflated.

Length about 140 mm. (5.50 inches); tail vertebræ 27 (1.05); hind foot 17.5 (.70).

Type locality, Pigeon Spring, Mount Magruder, Nevada.

White and Inyo Mountains, California and mountains of western Nevada, principally in sagebrush in dry, barren localities.

Subgenus **Chilotus**.

Pelage comparatively short and dense; soles with five tubercles; pattern of surface of lower molar without closed triangles; third upper molar with two or three closed triangles and six salient angles; skull low and flat with long slender rostrum.

Microtus oregoni BACHMAN. (Of Oregon.)
OREGON MEADOW MOUSE.

Above mixed bistre and blackish; below dusky washed with dull buff; feet dusky; tail blackish, slightly lighter below; ears blackish, longer than the surrounding pelage.

Length about 140 mm. (5.50 inches); tail vertebræ 42 (1.65); hind foot 17 (.67).

Type locality. Astoria, Oregon.

Oregon Meadow-Mice frequent dry open ground under cover of grass, or of logs in open forest from Humboldt Bay to Puget Sound.

Genus **Fiber** CUVIER. (Beaver.)

Size, largest of the family; skull strong, angular, very narrow between the orbits; molars rooted; basal part of lower incisor passing on tongue side of the first and second molars and on the outer side of third; parietals and interparietal very small; hind feet large, partly webbed, capable of being turned obliquely in swimming; tail long, widened perpendicularly and fringed with stiff hairs on the edges, the sides being nearly bare; underfur dense.

Fiber zibethecus pallidus MEARNS. (Civet like; pallid.)
PALE MUSKRAT.

General color above light glossy chestnut; sides russet; lower parts grayer; underfur light plumbeous.

Length about 445 mm. (17.50 inches); tail vertebræ 195 (7.70); hind foot 68 (2.67); ear from crown 16.5 (.65).

Type locality, Fort Verde, Arizona.

Pale Muskrats live along the Colorado River and its tributaries, but are not plentiful. They live in the banks of the main river and also in the banks of ponds and old channels containing still water. Muskrats have been reported from Carson River in Nevada, and they may cross the State line into the few suitable places in the upper part of the valley. I am quite sure that I have seen a reference to their occurrence in the Sacramento Valley, but I am unable to find it or recall the particulars.

I found Pale Muskrats in a small lake above Needles, on the Arizona side, but they were very few in number. On the Califor-

nian side of the Colorado River, a few miles below Ehrenberg,
Arizona I found a colony inhabiting a "slough". I trapped for
them unsuccessfully, but succeeded in shooting three by moonlight
and one after sunrise as they were swimming among the rules.
They were much smaller than the Muskrats that I used to trap in
the Mississippi Valley. Two weighed twenty ounces each. The
fur was thin and short, as might be expected in that warm climate.

I have seen no "houses" and can learn of none in the west.
These mounds of dead vegetation are common in the sloughs and
ponds of the northeastern States. The food of Muskrats generally is the stems and roots of aquatic plants. Fresh-water mussels and fish are also eaten. Occasionally vegetables are taken
from gardens near streams that they frequent. A burrow opened
by Schott near Yuma contained screw beans.

Family **Geomyidæ** (The Pocket-Gophers.)

Body stout, thickset; head wide and blunt; eyes and ears small; mouth peculiar in having no lips, the large incisors projecting through the ordinary skin, which is haired behind them, the real mouth opening just in front of the premolars; cheek pouches large and opening externally, these pouches being purse-shaped infoldings of the loose skin of the neck, lined with short hairs, reaching back nearly to the shoulders and held in place by small muscles; legs very short and strong; feet large, with five toes each; claws of fore feet very large; tail about half as long as the head and body, scantily haired, the tip endowed with tactile nerves; skull large; lower jaw massive, strongly curved; incisors very long and stout; squamosal much expanded; mastoids restricted to the occiput.

This family contains nine genera and more than one hundred nominal species and subspecies; a considerable number of these will probably be dropped when the genus is critically studied as a whole. The distribution is temperate North America, exclusive of the Middle and New England States, Mexico and Central America. Most of the genera and many of the species are Mexican. But one genus is known to occur in the United States west of the Rocky Mountains.

The food is mostly vegetable, a large part consisting of roots and tubers. Succulent plants are drawn into the burrows and eaten. It is probable that such worms and insects as are incidentally found are also eaten. That part of the food obtained beneath the surface is found by the laborious process of digging burrows through the soil. Openings to the surface are made every few feet for the purpose of disposing of the soil excavated. If food plants chance to stand quite near to these openings they are cut and drawn into the burrow. If seen at a little distance from the burrow the animal prefers tunneling to them, rather than venture a few feet on the surface, so reluctant is the animal to expose itself by leaving its burrow. They are cautious but

not cowardly. They seem to fear nothing and will attack anything that molests them.

Being subterranean in habit, working in the dark, they are
active at all hours, but are least so in the middle of the day. In
soft earth the digging is done with the fore feet, but in hard
soil the incisors are used to loosen it. As the earth is loosened
it is scratched back to the hind feet which pass it on until enough
for a load is ready, when the animal turns around, brings the
wrists together under the chin, the fore feet extended outward, and then, propelled only by the hind feet, the dirt is pushed
ahead of the animal to the outer opening of the burrow, when
the dirt is thrown out by a quick flirt. They run backward
nearly as rapidly and easily as forward, the sensitive tip of the
tail being used as a guide.

The pelage of the adult is commonly somewhat different
from that of the young, and in some species there is also seasonal
changes. They breed pretty much throughout the year or
through the warm months in the colder part of their habitat,
but it is not known whether the females breed more than once
a year. The young are born in an undeveloped condition. Two
to six constitute a litter.

Genus **Thomomys** Maximilian. (Heap—mouse.)

Front surface of incisor without a longitudinal groove, or
but a small one very near the inner edge; upper and lower molars
with two enamel plates, one anterior, one posterior; external ears
evident though small; four pairs of mammæ in most species.

Dental formula, I, 1—1; C, 0—0; P, 1—1; M, 3—3×2=20.

Thomomys fulvus nigricans Rhoads. (Fulvous; blackish.)

TAWNEY POCKET-GOPHER.

Variable in color; above usually yellowish bistre mixed with
black dorsally from crown to hips; sides lighter; below grayish

white tinged with buff; the slaty bases of the hairs showing through more or less according to the amount of wear; feet and tail pale buffy gray. A proportion are much darker, clove brown or dark sepia, and usually larger than the average. Others are redder, tawny, cinnamon or russet, and smaller than the average. The skull is small, light and comparatively smooth; rostrum broad and rather short; nasals long and narrow, projecting as far forward as the incisors; interparietal rectangular or pentagonal; temporal ridges small and wide apart except in aged animals; zygomata widest anteriorly in fully adult animals; groove near inner edge of upper incisor small but usually distinct.

Length about 200 mm. (7.87 inches); tail vertebræ 66 (2.60); hind foot 27 (1.06); ear from crown 6 (.24). Weight three to five ounces.

Type locality, Witch Creek, San Diego County, Califorina.

Abundant in the mountains and foothills of southern California. Less common in the mesas and valleys.

Thomomys monticolus ALLEN. (Mountain—inhabiting.)
MOUNTAIN POCKET-GOPHER.

Pelage long and soft; ears long; above fawn color or mars brown with a silvery gloss; sides and lower parts buff, the plumbeous bases of the hairs showing through; feet and tail pale buff; skull similar to that of *fulvus nigricans,* but with nasals shorter and wider anteriorly; zygomata widest posteriorly; interparietal broadly petagonal.

Length about 195 mm. (7.70 inches); tail vertebræ 66 (2.70); hind foot 26 (1.03).

Type locality, Mount Tallac, Eldorado County, California.

Common in the northern Sierra Nevada and throughout the northeastern part of California, from about 4,000 feet altitude nearly to timberline.

Thomomys monticolus pinetorum MERRIAM. (Of the pines.)

PINE-WOODS POCKET-GOPHER.

Similar to *monticolus* but smaller, skull shorter with broader zygomata; color paler; above pale fulvous; nose dusky.

Type locality, Sisson, Siskiyou County, California.

Common around the base of Mount Shasta, grading gradually into *monticolus* on the higher parts of the mountain.

Thomomys alpinus MERRIAM. (Alpine.)

ALPINE POCKET-GOPHER.

Similar to *fulvus*. *Light pelage;* above sepia or drab brown suffused with fulvous; below plumbeous washed with ochraceous buff. *Dark Pelage;* plumbeous tipped with russet brown. Skull small, rounded; nasals rather short; zygomata wide.

Length about 220 mm. (8.65 inches); tail vertebræ 63 (2.50); hind foot 30 (1.20).

Type locality, Cottonwood Meadows, Mount Whitney, California.

Common in the high southern Sierra Nevada.

Thomomys perpallidus MERRIAM. (Very pale.)

PALLID POCKET-GOPHER.

Very pale; varying (principally with locality) from yellowish drab to ochraceous buff, cream buff or grayish white; below dull white, the hairs sometimes white to the roots but more often pale plumbeous basally; mouth parts more or less brown; skull rather large, smooth; rostrum rather wide; nasals long, depressed, rather narrow, squarish posteriorly; frontal flat, often slightly concave; interparietal about as long as wide, the front outline subcircular.

Length about 233 mm. (9.15 inches); tail vertebræ 85 (3.35); hind foot 33 (1.30); ear from crown 6 (.24).

Type locality, Palm Spring in the northwestern corner of the Colorado Desert, California.

The Pallid Pocket-Gopher is found in the arid Colorado and Mojave Deserts. It is common in a few localities, but from the barren nature of this region it is necessarily rare in many parts of these Deserts.

Thomomys perpallidus perpes MERRIAM.
GOLDEN POCKET-GOPHER.

Above varying from yellowish drab to grayish ochraceous buff; face ashy; below grayish white; throat white; mouth parts ashy or plumbeous; feet and tail grayish white; skull somewhat smaller than that of *perpallidus;* rostrum smaller; interparietal squarish.

Length about 212 mm. (8.25 inches); tail vertebræ 66 (2.67); hind foot 29 (1.15).

Type locality, Lone Pine, Inyo County, California.

Owen Valley and the western part of the Mojave Desert.

Thomomys bottæ EDOUX and GERVAIS.
CALIFORNIA POCKET-GOPHER.

Above sepia mixed with black; sides paler; below slaty tipped with cinnamon or ochraceous buff; mouth and nose blackish except around the incisors where it is white; lining of pockets often white, but their edges blackish; feet dull white; tail dusky above basally, the remainder whitish. *Young;* paler and tinged with fulvous. Skull massive, angular: rostrum short and nar-

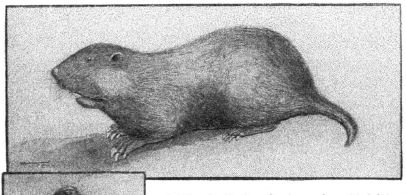

California Pocket Gopher.   One-third life size.

row; nasals short; incisors projecting forward. their front surfaces paler than usual in this genus; zygomata broad. widest posteriorly; interparietel small; triangular, narrowed and nearly overgrown by the temporal ridges of old age; occiput truncated posteriorly.

Length about 240 mm. (9.50 inches); tail vertebræ 75 (3); hind foot 32 (1.25).

Type locality, near Monterey, California.

Abundant in the coast region of central California.

The various species of Pocket-Gophers found in California (except in the deserts) are so much alike externally and in habits that the following account of their habits will apply to all.

Pocket-Gophers are thoroughly distributed throughout the

State wherever vegetation grows, from the seacoast to as high in the mountains as sufficient soil to work in occurs, except in land regularly subject to overflow. They are naturally most abundant in rich loose soils.

The food is principally the roots and succulent stems of plants, such as garden vegetables generally, potatoes, alfalfa, etc., as well as very many species of wild plants. The roots of fruit trees are often eaten, though in large trees but a portion of the the roots of any particular tree is eaten and the ill effects are not as noticeable as with young trees.

The burrows or runs are commonly less than a foot below the surface, but vary with soil and season; as the object in digging the run is to find food it is naturally dug at the depth where roots are most abundant. These runs are practically endless as they are being extended daily, except perhaps in the dry season, when comparatively little new work is done on account of the hardness of the soil. In very few localities in this State are the Gophers hindered by frozen soil, but in such places they work deeper, or occasionally on the surface under the snow, these surface runs being often filled with earth later, becoming very noticeable after the snow has melted. Openings to the surface are made at varying intervals for the purpose of getting rid of the soil excavated in making the runs, the dirt being thrown out in mounds containing a quart to a peck of earth. When the run has been excavated an inconvenient distance beyond the last opening that is closed and a new one made. These openings made for the purpose of carrying out the loosened earth are started at the side of the main run, gradually turning upward, and come to the surface one or two feet at one side of the main run. When abandoned these side runs are often packed full of earth.

The Gophers pass back and forth several times a day over the newer part of the main run, probably spending their hours of repose some distance from the new end of the burrow. In some seasons they make a nest of dry grass, but in the warmer part of the year they apparently use none, but lie down wherever they happen to be. They are more or less active at all hours, but much less earth is thrown out during the hours of bright sunshine.

No doubt the female makes a warm nest for her young, but I have never happened to find such a home, and it is probably deeper in the ground than the main run and ordinary sleeping nests. I have taken females suckling young at all times of the year. It is not known whether they have more than one litter annually or not, but it is probable that they do. Two to six young constitute a litter. I find in my note book a note of having taken a female containing six fœtuses, on March tenth.

It is seldom that more than one Gopher inhabits a run. At times a pair may be found inhabiting a run, but not often. The young shift for themselves before they are half grown. These young are very difficult to trap because of their small size. I have taken them from runs that would hardly admit my thumb. In such cases several may be found in a quite small area. No doubt the young commence work from the nest and gradually drift apart.

Many people suppose that the earth thrown out is carried in the cheek pouches by the Pocket Gophers, but such is not the case, the pouches being used only for carrying food. They seem to prefer to carry the food some distance back in the run to eat it in quiet. Small bits of food are pocketed as found and work is continued until sufficient is gotten to be worth while stopping to eat or store away. Considerable quantities of food are stored for future use, though not to as great an extent here as in colder climates. If food is seen or smelled on the surface at a little distance from the opening of the run the Gopher prefers running a tunnel to it rather than to venture far on the surface. The little fellows are good engineers, for I have many times seen pumpkins and melons eaten to a

shell with no sign of the run beneath until the shell is picked
up or rolled over, and no mound within several feet.

The strong incisor teeth are apparently used in hard soil in
loosening the earth, but most of the digging is done with the
fore feet. When sufficient soil is loosened and thrown behind
it the Gopher turns around, brings the wrists together beneath
the chin with the palms in front and the claws outward behind the pile of loosened earth and pushes the pile before it,
using the hind feet only in propulsion. The animal can push
a much larger amount before it than it could carry in its pockets.
On reaching the surface a flip of the fore feet throws the earth
a little distance, the action appearing as if the earth was thrown
out of the pockets with the fore feet.

Sometimes the Pocket-Gophers run backward instead of
turning around in the run, especially if the run happens to
be in hard earth and narrow. Dr. Merriam kept a live Gopher
in captivity to study its habits, and found that it could run
backward easily, and nearly as rapidly as forward. The nearly
naked tail is used as a feeler, and is quite sensitive as an organ of
touch. The hearing appears to be fairly good, but the sight
is poor. Smell is probably the principal sense in locating food.

Pocket-Gophers are sharp tempered animals and very courageous. They do not hesitate to attack anything that inter-
feres with them, and the bite of a trapped Gopher is sufficiently
severe to be dreaded. From the circumstance of their living
alone one may surmise that they are surly and quarrelsome.

Pocket-Gophers are a serious pest to the farmer and fruit
grower. Having had considerable experience in trapping them
I may be able to give some useful hints to those who may wish
to try to get rid of them. Poison is not as useful in the case
of Gophers as it is with some other pests, as it is likely to be
pushed out of the run instead of being eaten. Poisoned grain
can be used, or a little crushed strychnine in a raisin or a bit
of apple or potato. If any of these be used place it in the main
run after clearing out all lumps of dirt and close the run thor-

oughly. It will thus be more apt to be eaten. A "smoker"
has been advocated and sold for the purpose of suffocating
Gophers. This implement works well with ground squirrels
but is not often effective with Gophers on account of the great
length of the runs and the difficulty of forcing the smoke far
enough. Bi-sulphide of carbon is moderately effective with
Gophers. To use it pour two tablespoonfuls on a bunch of rags,
waste or cotton and place it in the main run where fresh mounds
show the recent presence of the animal. Close the opening thoroughly to retain the fumes. Remember that the fumes are ex-
plosive. The most effective of all methods where it is practicable is drowning by flooding with water when thoroughly done.
For many people the main reliance must be on traps.

Trapping is most effective in the rainy season. The soil
is then in the best condition to work and the Gophers are more
active and less suspicious. By persistent trapping a place may
be cleared of Gophers; afterward one must continually watch
around the borders of the place to catch the immigrants as they
begin to work in. Neighbors joining in the work can make
it most effective. Two or three kinds of traps are needed for
different conditions in trapping. Some persons prefer one style,
others another, there is considerable choice and several efficient
styles. I prefer the "C V" trap supplemented with the common
steel trap, Newhouse pattern, No. O size. The small size of
"C V" is most useful as the large size is too large for most
runs. If cats have the run of the place the traps will need
staking or they may be carried away; a strong cord can be used
on the "C V" traps. The "C V" traps as usually made have
the triggers too far from the entrance. If they are of the
pattern having sheet iron triggers set the trap and push the
trigger toward the entrance, bending it considerably, so that the
trigger will be pushed before the Gopher gets so far through
the entrance. The other implements necessary are a shovel, a
bit of heavy hoop iron bent in a fish-hook shape for widening
the opening and drawing out the earth, some bits of board for

closing the openings, and, if you have many traps out, stakes to mark the places where the traps are set.

The rounds should be made twice a day, early in the morning and at night, as the Gophers are then actively at work. Look for fresh mounds. If an open hole is found widen it sufficiently to insert a "C V" its full length. Leave that hole open as the Gopher will be back in a few minutes to close it. If a fresh mound is found with the exit closed use the shovel carefully. If the run is not readily found it may perhaps be found by feeling with the end of the hoop iron or the finger, as the earth in the side run should be softer than its surroundings. Try to find the side run without breaking into the main run. If you find it, set a "C V" in it if there is room out side of the main run, and nearly close the run with a piece of board, leaving a little light to tempt the Gopher to close the hole. If there is not room enough in the side run to use a "C V" set a steel trap in the main run with the pan and jaws level with the floor of the run. See that no lumps of dirt are left in the run to give the Gopher warning. Close the side run thoroughly in this case as you are trying to get the Gopher as he makes the rounds of his run as usual. If you chance to open the main run at a "C V" in each branch and close the openings tight behind them. The Gophers will fill a considerable proportion of the traps with earth and fail to spring them. Reset the traps and close the hole. When traps are set in the main run the Gophers are more likely to fill them and make a new run a few inches behind the old run. Now and then a Gopher will prove to be cunning and difficult to catch. After trying such an animal a few times change to some other form of trap. It is seldom that a Gopher is too smart to be caught, but I have been baffled a few times; commonly they are careless and easily caught.

A good cat that will hunt Gophers is valuable, but the most efficient helper that a farmer can have is a pair of the much persecuted barn owls. These birds live principally on Gophers, and during the breeding season a pair will catch a dozen each night. Weasels are very fond of Gophers and are able to follow the runs of adults. Bull snakes or gopher snakes feed principally on gophers. They have no venom and should never be destroyed.

Thomomys bottæ pallescens RHOADS. (Pale.)
SOUTHERN POCKET-GOPHER.

Paler and more tawny than *bottæ*; an indistinct dusky dorsal stripe; sides usually with an indistinct ochraceous buff stripe separating the color of the upper parts from that of the belly; averaging larger than *bottæ*; skull similar; rostrum broader; incisors heavier and their front surfaces deeper yellow.

Type locality, San Bernardino Valley, California.

The Southern Pocket-Gopher is abundant in southern California along the coast and in the valleys and is more or less common in the mountains.

One April morning I had an opportunity to watch a Southern Pocket-Gopher at work. It was wary but not shy. It saw and watched me several seconds at a time. It paid a little attention to vocal sounds that I made but not much. It seemed to try to scent me. The light breeze blew toward it, distance ten feet. When first noticed I think its pockets were empty. I saw it gather some plants, including young wild oats. The transfers of plants from mouth to pockets were made very quickly, but I could not see just how it was done as its back was toward me, though I could see the pockets swell. It went down and brought more earth after a few seconds disappearance, repeating this several times, occasionally picking more "greens." That is, it did not immediately go off and eat or cache its food, but worked on with the food in its pockets, occasionally adding to its amount. In coming out of the run to gather the plants it did not walk at its full height, but crouched, dragging its belly on the ground, the hips and shoulders showing prominently above the vertebral outline. In pushing out the earth before it the nose was kept raised over it, not buried in the earth. The retreat was almost instantaneous after the earth was flirted away.

The following Pocket Gophers I have not seen. I give a summary of the original descriptions.

Thomomys laticeps BAIRD.
BROAD-HEADED POCKET-GOPHER.

Above yellowish brown, blackish dorsally; below tinged with reddish; tail about half as long as head and body; feet large; skull very broad; rostrum short.

Length 178 mm. (7.65 inches); tail vertebræ 64 (2.50); hind foot 27 (1.06).

Type locality, near Humboldt Bay, California.

Thomomys leucodon navus MERRIAM.
RED BLUFF POCKET-GOPHER.

Above fulvous brown; below buff ochraceous; skull small but very strong and ivory-like in texture; zygomata broadest posteriorly; nasals cuneate, usually notched behind; incisors projecting forward, their faces yellow.

Length 196 mm. (7.75 inches); tail vertebræ 65 (2.55); hind foot 27 (1.06).

Type locality, Red Bluff, California.

Thomomys angularis MERRIAM.
SAN JOAQUIN POCKET-GOPHER.

Above fulvous grizzled with black; below plumbeous strongly washed with buffy ochraceous; feet and tail whitish; skull large and massive; braincase broad; nasals emarginate posteriorly; interorbital region rounded.

Length 255 mm. (10 inches); tail vertebræ 75 (3); hind foot 32 (1.25).

Type locality, Los Banos, Merced County, California.

Thomomys angularis pascalis Merriam.
FRESNO POCKET-GOPHER.

Similar to *angularis* but smaller; above more buffy yellow; below very much paler and often marbled with patches of white; skull smaller and smoother.

Length 210 mm. (8.25 inches); tail vertebræ 70 (2.75); hind foot 30 (1.18).

Type locality, Fresno, California.

Thomomys operarius Merriam.
OWEN VALLEY POCKET-GOPHER

Above buffy yellowish or buff gray; below plumbeous strongly washed with white; skull short, broad and massive; rostrum short and broad; interorbital region broad; temporal ridges well marked.

Length 217 mm. (8.50 inches); tail vertebræ 67 (2.65); hind foot 29 (1.15).

Type locality, Keeler, Inyo County, California.

Thomomys cabazonæ Merriam.
CABEZON POCKET-GOPHER.

Above varying from buffy ochraceous to dull drab brown; below whitish or pale salmon; skull small, angular; zygomata broadest anteriorly; interparietal rectangular; nasals long.

Length 220 mm. (8.25 inches); tail vertebræ 78 (3.06); hind foot 30 (1.15).

Type locality, Cabezon, San Gorgonio Pass, Riverside County, California.

Thomomys fuscus fisheri Merriam.
FISHER POCKET-GOPHER.

Similar to *fuscus* but very much paler, grayish brown instead of dull fulvous brown; skull similar to that of *fuscus* but shorter; zygomata more squarely spreading; premaxillæ shorter and broader posteriorly; bullæ less swollen; incisors narrower.

Length 192 mm. (7.60 inches); tail vertebræ 58 (2.30); hind foot 25 (1).

Type locality, Beckwith, Plumas County, California.

Family **Heteromyidæ.** Pocket-Rats and Pocket-Mice.

Cheek pockets large and opening externally, similar to those of *Geomyidæ*; fore legs of moderate length; hind legs more or less lengthened; tail usually as long as or longer than head and body; skull thin and smooth; rostrum long and tapering; nasals long, projecting beyond the incisors and semi-tubular anteriorly; frontals wide; no inteorbital foramen but a perforation on the sides of the maxillary instead; occipital region formed mostly of the mastoids; temporal region inflated, sometimes enormously; zygomatic arches very slender, depressed; lower jaw small and weak; coronoid process very small; angular process twisted obliquely.

Dental formula, I, 1—1; C, 0—0; P, 1—1; M, 9—3×2=20.

This is a small family of seven genera, divided in two subfamilies. The family is American, Mexico being apparently the center of distribution, none of the family being found east of the Mississippi River. It has been very imperfectly known until recently and has many interesting peculiarities. A singular characteristic of Pocket-Rats and Pocket-Mice is their ability to go without water and if necessary without eating moist food. Most species inhabit arid regions or deserts, though a few species are found in regions of moderate rainfall, provided the climate is comparatively warm. They do not endure cold well, very few being able to live in localities where the ground freezes too hard to plow.

The food is principally seeds, but leaves and stems of plants are occasionally eaten. Seeds are commonly stored in chambers of burrows or sometimes in surface caches. In some of the Californian valleys harm is done by members of this family through carrying off and hiding grain, though it is seldom done to a noticeable extent. Yet the total loss to grain growers must annually amount to a considerable sum, because of the abundance and industry of these little animals. I am not aware of any other harm being done by them and there is some compensation in the quantity of weed seeds destroyed. Their mouths are small; the largest food I have known them to eat is acorns.

The family is digitigrade; nocturnal; terrestrial and sub-terranean, living in burrows but gathering their food from the surface. There is little if any seasonal change of pelage. The sexes are alike, but the young sometimes differ from the adults.

Subfamily **Dipodomyinæ**

Skull triangular in general outline; molars rootless; temporal region, enormously inflated; interparietal small, narrowed from the sides, sometimes obliterated; hind legs lengthened; size comparatively large.

Genus **Perodipus** Fitzinger. (Pouch—two-footed.)

Upper incisors grooved in front; temporal region greatly inflated; supraoccipetal, interparietal and parietals greatly reduced in area; zygomatic arch expanded to a large thin plate in front of the orbit; hind feet with five toes, the inner toe minute and situated higher than usual; soles heavily haired; tail longer than head and body, four striped; eyes large; ears large, rounded, the front border inflexed; pelage soft, without spines.

The pattern of coloration of all the species of *Perodipus* and *Dipodomys* is the same. There is a dark patch at the base of the whiskers, a white spot over the eye, another below or behind the ear, a white stripe across the thighs to the base of the tail, a dark stripe on the upper side of the tail, usually another on the under side leaving the sides of the tail white, fore feet and all the lower parts from mouth to tail white to the roots of the hairs.

The beginner will find the species of *Perodipus* very puzzl-ing, and will probably be unable to assign his specimens satis-factorily. The division of this genus has been carried to an unnecessary refinement. I have omitted several nominal species or subspecies rather than add to the difficulty. *Dipodomys* is not quite so difficult, as size and color are more diversified.

Perodipus agilis GAMBEL. (Nimble.)
GAMBEL POCKET RAT.

Above yellowish bistre mixed with black, the basal half or three-fourths of the hairs slate gray; sides ochraceous buff; tail crested, the hairs toward the end being lengthened, principally on the upper side; upper tail stripe as dark as the back, the lower stripe but little lighter and continuous to the end, but tip usually with white preponderating; soles of hind feet blackish; ankles dull black posteriorly; ears large; skull narrow; supra-occipetal very narrow; interparietal narrow; nasals narrow, the outer edges of the posterior half parallel; maxillary arches comparitively narrow. *Young;* darker, more slaty; hairs of terminal part of tail not lengthened.

Length about 288 mm. (11.33 inches); tail vertebræ 180 (7.10) hind foot 42 (1.65); ear from crown 14 (.55).

Type locality, Los Angeles, California.

Gambel Pocket-Rats are common in the coast region of southern California and on the sides of the mountains to 3,000 feet altitude or higher. They are not often found in brush, or in rocky ground, preferng open valleys having a good growth of annual plants, the seeds and leaves of these plants forming the principal part of their food. Grain is sometimes stored in their burrows, but grain land, especially if summer-fallowed, affords too little subsistence in the dry part of the year, and is usually deserted for places where seed producing plants remain on the ground all the year. Occasionally the borders of grain lands are invaded, but the depredations of Pocket-Rats are rarely serious, and these are partly balanced by the large amount of weed seeds eaten.

The following notes on an opened burrow of this species are given to illustrate some of their habits, which are similar to those of Pocket-Rats in general. I had noticed the entrance to a burrow at the side of a path a few yards from the kitchen door of my house; the burrow had been used some months, still the house cat (a very good mouser) had not caught the oc-

cupant. One day in January I set a box trap near the burrow
and the next morning found a Gambel Pocket-Rat in it. This
animal I kept alive in a box. The second evening following,
I found my dog playing with a young Pocket-Rat near the mouth
of the burrow. The next day I dug open the burrow to see
what its internal arrangement was. The burrow was oval in
section, the perpendicular diameter greatest, being a little more
than two inches. The entrance sloped gently downward, and the
main burrow was about eight inches below the surface for about
eight feet, then about a foot and a half below for six feet, where
it terminated in another entrance which I had not previously
noticed, as it was under a small perennial plant. This last entrance was nearly perpendicular for six or seven inches. There
were half a dozen branches to the burrow, varying from a few
inches to three feet in length, each terminating in a chamber of
greater diameter than the burrow.

In one chamber was the nest, a mass of nearly a quart of
the hulls of grass seeds. The other chambers were used as granaries. These contained acorns, seeds of poverty grass, and of
chrysanthemums and other flowers from a bed in front of the
house. Most of the granaries were closed with earth. The open
one was but part full and probably was the one from which food
was then being used. The various granaries contained respectively 149, 27, 24, 131, 139 and 26 acorns. The weight of the
acorns and seed was forty two ounces; that of the adult female
was two and a quarter ounces.

In the burrow, a foot or more from the nest was another
quite young Pocket-Rat. It seemed to very cold and hungry
and made a grating squeaking sound, which it kept up some
time after being put with its mother. I could not see that it
suckled, and think it did not, though it crept under its mother
and persisted in staying there. It ate shortly after being put
with its mother.

The acorns in the granaries were brought from a tree standing more than a hundred feet from the burrow. On trial an

acorn was found to slip easily into a cheek pouch, the tip of the acorn projecting outside. The partly eaten acorns were commenced from the base, probably because that part of the shell covered by the cup was thinner and more easily bitten through.

There was no pile of earth at either entrance and the upper entrance, under the plant, was closed with earth. The absence of a pile of earth at the entrance was not unusual, but more often the pile of earth is present. The entrances of burrows are most frequently closed in the daytime. This burrow was longer

Gambel Pocket Rat.   Nearly one-half life size.

than usual, and probably contained a greater amount of food than is usual.

The gait is a series of leaps; if hurried these are very rapid and three or four feet in length. As they can turn very abruptly, it is very difficult for a dog to catch one. In leaping from a position of rest they can go off as if a spring in them had been suddenly released. They cannot continue a rapid run far; the longest that I remember seeing in the daytime was about fifty yards, when the animal entered its burrow. I have several times seen Pocket-Rats abroad in the daytime, but this is not a common habit.

The only vocal sounds that I have heard from adults are squeaks of pain when caught in a trap or otherwise hurt. The greater number of young are born in the spring, but reproduction continues to some extent through the summer and autumn. Three to five young at a birth appear to be the usual number.

Pocket-Rats make interesting pets. They do not resent being handled, though not really liking it. They do not often try to escape when held in the hands. They require considerable provocation and rough usage before attempting to bite. As they can open the mouth but a short distance and the upper incisors are bent backwards they are unable to bite deeply, scarcely more than through the skin. They keep in good health in captivity on grain alone. They do not bear cold well.

Perodipus ingens MERRIAM.
BIG POCKET-RAT.

Size large; above buffy ochraceous; tail with upper and lower stripes black and a white pencil; ears small; skull very large and massive.

Length about 350 mm. (13.75 inches); tail vertebræ 190 (7.50); hind foot 52 (2.05).

Type locality, Painted Rock, San Luis Obispo County, California.

Known only from the Carrizo Plain.

Perodipus venustus MERRIAM. (Beautiful.)
SANTA CRUZ POCKET-RAT.

Similar to *agilis,* but very much darker; top of head, back, and thigh patches dusky, finely grizzled with ochraceous; hairs of rump forming a black patch just in front of basal white ring of tail.

Length about 315 mm. (12.40 inches); tail vertebræ 190 (7.50); hind foot 45 (1.77).

Type locality, Santa Cruz, California.

Santa Cruz and Santa Lucia Mountains.

Perodipus goldmani MERRIAM. (For L. J. Goldman.)
GOLDMAN POCKET-RAT.

Darker than *agilis* and lighter than *venustus;* bases of hairs
of upper parts slaty black; ears smaller than *agilis;* dark tail
stripes becoming lighter colored toward tip; skull differing from
that of *agilis,* wider, particularly across the maxillary arches;
nasals narrowed posteriorly and slightly constricted in the middle, but considerably wider in the anterior third; supraoccipetal
very narrow; interparietal small and very narrow.

Length about 313 mm. (12.30 inches); tail vertebræ 185
(7.30); hind foot 45 (1.77).

Type locality, Salinas, Monterey County, California.

Salinas Valley and other valleys of that general region.

Perodipus panamintus MERRIAM. (Of the Panamint
Mountains.)
PANAMINT POCKET-RAT.

Similar to *agilis;* lighter colored with much less black intermixed; ears small; the lower tail stripe becomes obsolete on
the terminal third; soles gray; skull similar to that of *goldmani*
but anterior third of nasals narrower; supraoccipetal wider; interparietal wider, its width half to two-thirds its length.

Length about 300 mm. (11.80 inches); tail vertebræ 180
(7.10); hind foot 45 (1.77).

Type locality, Panamint Mountains, California.

Panamint Pocket-Rats are found throughout the Mojave
Desert region, though they are common in but few places.

Perodipus streatori MERRIAM. (For C. P. Streator.)
STREATOR POCKET-RAT.

Similar to *agilis* but larger; ears smaller; top of tail white;
skull larger and heavier; fronto-parietal suture strongly convex
forward in the middle.

Length about 295 mm. (11.60 inches); tail vertebræ 180·
(7.10); hind foot 43 (1.70).

Type locality, Carbondale, Mariposa County, California. Western foothills of the Sierra Nevada.

Perodipus microps MERRIAM. (Minute—like.)
INYO POCKET-RAT.

Small; ears small; above pale buffy ochraceous; skull small, narrow, with narrow braincase.

Length about 270 mm. (10.60 inches); tail vertebræ 158 (6.20); hind foot 41 (1.62).

Type locality, Lone Pine, Inyo County, California. Northern part of the Mojave Desert.

Genus **Dipodomys** GRAY. (Two-footed—mouse.)

Hind foot with but four toes, otherwise similar to *Perodipus,* but varying more in color and size. See under *Perodipus* for color pattern.

Dipodomys californicus MERRIAM.
CALIFORNIA POCKET-RAT.

Very similar to *Perodipus goldmani* in color and size; skull similar but narrower interorbitally, nasals narrower; supraoccipetal wider; mastoids and audital bullæ much smaller.

Length about 300 mm. (11.80 inches); tail vertebræ 185 (7.80); hind foot 43 (1.70).

California Pocket-Rats probably occur in all the northern coast Counties. I have taken them in Lake and Mendocino Counties. In the Pacific Railroad Reports, Dr. Suckley says that a "Kangaroo Rat" is common on the Salmon River; this is doubtless the present species. They inhabit brush and forests as well as open land, though probably not living in dense forests. I trapped one among redwoods near an open glade. I did not observe anything peculiar in their habits in other respects. They did not seem to be common. One trapped in an old vacant house near the end of April, contained three fœtuses.

Dipodomys californicus pallidulus BANGS.
COLUSA POCKET-RAT.

Similar to *californicus* but paler; above wood brown, shading to cinnamon on the sides; tail above sepia, beneath white.

Length 290 mm.   (11.40 inches); tail vertebræ 181 (7.15); hind foot 42 (1.65).

Type locality, Sites, Colusa County, California.

Dipodomys deserti STEPHENS.   (Of the desert.)
DESERT POCKET-RAT.

Large and pale; above grayish buff, the hairs ashy gray at base; shoulders and upper part of sides lighter, the hairs white at base; white spot behind the ear large, sometimes reaching the shoulder; soles of hind feet dirty white; upper side of tail white at base, then buffy or brownish to past the middle, then drab gray to the white tip, under side white; other marking as usual; skull large; inflation of mastoids extreme; supraoccipetal narrowed almost to a line on the upper surface; interparietal usually obliterated in adults.

Length about 340 mm.   (13.40 inches); tail vertebræ 205 (6.10); hind foot 53 (2.10); ear from crown 15 (.60).

Type locality, Mojave River, near Hesperia, California.

Desert Pocket-Rats occur in southern California, southern Nevada, western Arizona, northwestern Sonora and northeastern Lower California.   The northwestern extreme of their range is Owen Valley, Inyo County, where I have trapped them near Alvord.   None have been taken on the coast side of the mountains.

This species occurs in small colonies, less commonly in pairs, rarely singly.   The habitation is often a labyrinth of intercommunicating burrows from a few inches to two feet beneath the surface, commonly under a low mound formed of sand and dust drifted about a shrub, but sometimes in a level space.   Frequently the interior is honeycombed with burrows until little more than a surface shell remains and one breaks through on stepping there. A horse soon learns to avoid these burrows, walking being sufficiently tiresome in this sandy region without falling into rat-nests. Not all the burrows are excavated to this extent however, many being simple burrows a few feet in length with two entrances, or with a branch or two.

In places where much camping is done, such as by springs on the road from one mining camp to another, the Pocket-Rats are in the habit of coming about camp at night to pick up grain scattered by the horses and other food, becoming comparatively tame, as no one harms them. I never knew a dog to catch one, as they get under way very quickly, and vanish in the nearest burrow; in such places they have many burrows, perhaps for just such emergencies.

I have kept Desert Pocket-Rats in captivity several times, at one time having two for several months. Some of the habits of these as observed in captivity are worth recording. The first I kept was caught at the type locality in November. At that season food had become scarce and they were hungry and easily trapped. The box trap was set a few yards from camp; hearing the door fall I immediately took the animal out and put it in a cage and put some grain in the cage. It was amusing to see the eagerness with which it began filling its pockets. It stuffed them so full that it must have been almost painful. It would not stop to eat, but hunted about for some exit; not finding one it ejected the contents of its pockets in a corner out of the firelight and went back for more. This time it ate a little grain, but soon gathered the remainder and deposited it with the first. After eating a little more it refilled its pockets and hunted about for a better place to make a cache, seeming to think its first choice insecure.

On arriving home I put a little dry earth in the cage. This pleased the Pocket-Rats and they enjoyed a good dust bath, rolling in the earth and pushing along on their bellies. They looked much better for their dust bath, the roughened pelage becoming smooth and glossy.

None of my captive Pocket-Rats would drink water. For food they preferred grain, but also ate such vegetables as sweet potatoes and the leaves of beets and cabbages. They consumed little more than a heaping tablespoonful of wheat or barley in twenty-four hours. The two pockets together held a heaping tablespoonful of grain and therefore would carry nearly a full day's ration.

The pockets are filled by the fore feet used as hands. The filling is done so rapidly that when a hard grain, like wheat, is used a continuous rattling sound is made. The ejection of the grain is aided by a forward squeezing motion of the fore feet, each foot making two or three quick forward passes scarcely occupying a second of time.

The position at rest was a curious one. At first the animal stood on all four feet, with the entire sole of the hind foot resting on the ground, some of the weight coming on the fore feet. Presently the hind feet would hitch forward until the center of gravity came over the hind feet, thus taking all the weight, then often the fore part of the body would be raised slightly and the fore feet drawn up against the body. If disposed to sleep the bright eyes would slowly close, the fore feet droop until touching the ground, the nose come slowly down and backward until resting between the toes of the hind feet, and the now sleeping animal was nearly as round as a ball. This appears to be the common sleeping posture. If there be room the tail will be extended backward in a nearly straight line, but in cramped quarters it will be curved to one side or even alongside of the body, but in either case the basal part will be extended back far enough to give some support.

Dipodomys merriami simiolus RHOADS. (For Dr. C. Hart Merriam; a little mimic.)
MIMIC POCKET-RAT.

Very similar in color to *deserti*, but much smaller; tip of tail dark and a light brown stripe on under side of tail; soles of hind feet often partly bare.

Length about 240 mm. (9.50 inches); tail vertebræ 150 (5.90); hind foot 37 (1.45); ear from crown 10 (.40).

Type locality, Palm Spring (Agua Caliente), western end of the Colorado Desert, California.

Mimic Pocket-Rats are common in many parts of the Colorado and Mojave Deserts. A few occur on the northeastern slope of the mountains bordering the Colorado Desert, and in one place across the divide on the upper part of the Temecula River. These last are intermediate between *simiolus* and *parvus,* and perhaps should be classed with the latter subspecies. Those taken along the western border of the Colorado Desert are rather larger than the average elsewhere. They inhabit sandy land having a brief growth of annuals, the seeds of these being stored for use during the remainder of the year. In a few places where plants are plentiful these animals are common. They are crepuscular and nocturnal, not shy, often coming around the camp fire. They have several times run over me as I lay sleeping on the ground, and one even ran across my feet as I sat quietly by the camp fire alone. Their tracks in the morning show that they have gleaned thoroughly about camp the previous night for crumbs of bread, grain or other food. The young are born mostly in April and May, and are oftenest four in number.

Dipodomys merriami parvus RHOADS. (Little.)
SAN BERNARDINO POCKET-RAT.

Above broccoli brown, slightly tipped with black, the greater part of the hairs slate gray; sides light wood brown or clay color; soles of hind feet and leg above the heel blackish; upper and lower tail stripes distinct, blackish to end; side stripes white to mixed tip. Slightly smaller than *simiolus.*

Type locality, south side of San Bernardino Valley, California.

This subspecies appears to be rare. The few known specimens have been taken in Rêche Canon, a few miles south of San Bernardino. Their habits appear to be like those of the species in general.

Dipodomys merriami nitratus MERRIAM.
KEELER POCKET-RAT.

Similar to *simiolus;* averaging smaller; hind feet longer; dusky markings obsolete; hairs of back not tipped with black.

Type locality, Keeler, Inyo County, California.

Habitat, Owen Valley.

Dipodomys merriami nitratoides MERRIAM. (Resembling *nitratus.*)
TULARE POCKET-RAT.

"Similar to *nitratus* in size and and color but with strongly marked facial crescents meeting over bridge of nose; ears smaller."

Type locality, Tipton, Tulare County, California.

Dipodomys merriami exilis MERRIAM. (Smallest.)
LEAST POCKET-RAT.

Upper parts nearly uniform clay-color, darkened with sepia from abundant admixture of black tipped hairs, and darkest on the head; sides tinged with ochraceous buff; black crescents at base of whiskers sharply defined and meeting on the bridge of the nose; upper and lower tail stripes sooty blackish, meeting along terminal third, thus interrupting the white side stripes.

Length about 227 mm. (9 inches); tail vertebræ 136 (5.35); hind foot 35 (1.38).

Type locality, Fresno, California.

Genus **Microdipodops** MERRIAM. (Small—two footed-- like.)

General appearance similar to that of a large thickset

Perognathus with a large head and long hind feet; skull similar to *Dipodomys,* with the inflation of the temporal region carried to the farthest extreme known among mammals; zygomatic process similar to that of *Perognathus,* not widely expanded in front of orbit as in *Perodipus* and *Dcpodomys;* occipetal notch between the enormously inflated mastoids proportionally deeper than in any other genus of the family; supraoccipetal, interparietal and parietals greatly reduced; hind feet long; soles densely haired; five toes on each hind foot, the inner toe similar to that of *Perognathus* in size and location; tail indistinctly bicolor, somewhat longer than head and body; mammæ six, one pair pectoral and two pairs inguineal.

Microdipodops californicus Merriam.
CALIFORNIA DWARF POCKET-RAT.

Above grizzled yellowish olive; sides from nose to thighs cream buff, forming an indistinct stripe; below dull white to roots of hairs; no white stripe across the thigh; a dark crescent at base of whiskers; an indistinct white spot above the eye; feet grayish white; tail bicolor, dull buff below, buffy gray above, darkening toward the tip; whiskers mostly blackish, the longest reaching the shoulders.

Length about 160 mm. (6.30 inches); tail vertebræ 92 (3.60); hind foot 25 (1).

Type locality, Sierra Valley, Plumas County, California.

The above description (except measurements) is drawn up from a male that I caught in northwestern Nevada, five miles east of the California boundary. This may vary a little from the Plumas County animals. The only locality in California where I know of Dwarf Pocket-Rats having been taken is Sierra Valley, Plumas County. Animals of this genus have been taken in various localities in Nevada and in eastern Oregon. The two that I caught were taken in sandy land among sage brush, with grain baited mouse traps. I know nothing further of their habits.

Subfamily **Heteromyinæ**

Molars rooted; temporal region moderately inflated; inter-parietal large; hind legs not greatly lengthened; size small.

Genus **Perognathus** MAXIMILIAN. (Pouch—jaw.)

Upper incisors narrow, grooved in front; zygomatic process not greatly expanded in front of the orbit; tail about as long as head and body; hind feet with five toes, the inner toe small, situated but little above the other toes; soles nearly naked; pelage sometimes spiny on the rump and sides.

Subgenus **Perognathus**.

Mastoids comparatively large, projecting beyond the plane of the occiput; pelage soft, without spines or bristles.

Perognathus. panamintus MERRIAM. (Of the Panamint Mountains.)

PANAMINT POCKET-RAT.

Above grayish buff, often wth a pearly appearance caused by the pale buff ground color being overlaid by dark-tipped hairs; lateral line pale buff, not sharply defined; subauricular spot small and inconspicuous; fore legs buffy or white; under parts white; tail, above dusky, darkest toward tip, below buff or whitish; pelage long, full and silky.

Length about 143 mm. (5.63 ·inches); tail vertebræ 78 (3.07); hind foot 20 (.78).

Type locality, Perognathus Flat, (alt. 3,200), Panamint Mountains, California.

Panamint Mountains, California, eastward to St. George, Utah.

In April, 1891, Mr. Bailey and I found Panamint Pocket-Mice abundant in the type locality, which is a small sandy plain in the northern part of the Panamint Mountains, having a moderate growth of small shrubs and some grass.

Perognathus panamintus bangsi MEARNS. (For Outram Bangs.)
BANGS POCKET-MOUSE.

Similar to *panamintus* but paler; above pale vinaceous buff lightly mixed with black; lateral line blending with color of sides; under parts white; tail buffy white, slightly darker on upper side.

Length about 138 mm. (5.45 inches); tail vertebræ 80 (3.15); hind foot 19 (.75).

Type locality, Palm Spring, Riverside County, California.

Mohave Desert, Owen Valley and south to the foothills bordering the Colorado Desert on its southwestern side. Seldom common.

Perognathus panamintus arenicola STEPHENS. (Sand-inhabiting.)
SAND POCKET-MOUSE.

Similar to *bangsi* but paler and whiter; above cream buff, slightly mixed with blackish; no lateral line; mastoids greatly swollen and projecting much back of the occiput; interparietal small, its transverse diameter about equal the length.

Type locality, San Felipe Narrows, border of the Colorado Desert, California.

Sand Pocket-Mice inhabit the sandy gulches at the edge of the foothills bordering the Colorado Desert, sometimes being found a short distance out in the Desert. They appear to be rare. One morning about sunrise I found a little rattlesnake, of the species known as "sidewinder," with a dead Sand Pocket-Mouse in its mouth. The snake had crushed it and had just begun to swallow it, but disgorged on being struck with the butt of my gun.

Perognathus brevinasus OSGOOD. (Short—nose.)
SHORT-NOSED POCKET-MOUSE.

Above deep buff or grayish buff mixed with black, a band

(usually narrow) on the sides with but little intermixture of black; a small or obscure whitish spot at the base of the ear; a more or less distinct crescentic blackish line at the base of the whiskers; tail buffy, darker above. *Young;* drab buff above.

Length about 120 mm. (4.70 inches); tail vertebræ 62 (3.15); hind foot 18 (.70).

Type locality, San Bernardino, California.

Interior valleys of southern California. Occasionally common locally after wet seasons, but usually rare. I have never seen them abundant but once, this was after the wet spring of 1884, when they became plentiful at San Bernardino for two or three years, almost disappearing later.

The young are usually four in number, and are born in May, June and July. One female in the gray immature pelage taken June 4th, contained but two fœtuses. It is my impression that this species does not often dig burrows, but hides under weeds and dead leaves.

Perognathus pacificus MEARNS. (Of the Pacific.)
SAN DIEGO POCKET-MOUSE.

Very small; above ochraceous buff thickly mixed with black; a narrow lateral stripe of buff which widens on the sides of the head; distinct narrow black crescents at the base of the whiskers; a small white spot at the base of the ear and another indistinct larger buff one behind the ear; feet and lower parts white; tail buff below, darker above; skull small and narrow; mastoids less inflated than usual in the small species; transverse breadth of interparietal greater than the longitudinal.

Length about 109 mm. (4.30 inches); tail vertebræ 54 (2.12); hind foot 15.5 (.60); ear from crown 5 (.20).

Type locality, mouth of the Tijuana River, near the last boundary monument.

This exceedingly small Pocket-Mouse is one of the rarest of mammals yet, though some one may find them plentiful unexpectedly. Dr. Mearns and his assistant Mr. Holzner obtained three at the type locality in the extreme southwestern corner of San Diego County, while with the Boundary Survey in 1894. I had the good fortune to get the fourth, an adult female, in the northwestern corner of the same County, in Sept., 1903. These I believe to be the only specimens yet taken. I caught mine on a dry mesa a short distance back from the seashore.

Perognathus parvus mollipilosus Coues. (Little; soft —hair.)
COUES POCKET-MOUSE.

Above ochraceous buff thickly mixed with black; lateral line prominent; below white varying to tawny ochraceous on the belly; antitragus of ear prominently lobed; rostrum rather slender; mastoids but moderately developed; interparietal wide.

Length about 168 mm. (6.60 inches); tail vertebræ 88 (3.45); hind foot 22 (.86).

Type locality, old Fort Crook, Shasta County, California.

Mount Shasta and northeastern California. Coues Pocket-Mice have been taken on Mount Shasta at 7,800 altitude, which is unusually high for any Pocket-Mouse.

Perognathus parvus olivaceous Merriam.
(Shaded with olive color.)
GREAT BASIN POCKET-MOUSE.

Above ochraceous buff thinly mixed with black; lateral line buff; lower parts white, sometimes with plumbeous bases to the hairs; tail brownish above, white below; mastoids usually well developed.

Length about 178 mm. (7 inches); tail vertebræ 96 (3.75); hind foot 23 (.90); ear from crown 7.5 (.30).

Type locality, Kelton, Utah.

Great Basin Pocket-Mice range over most of the Great Basin, from northern Utah and southern Idaho west to the eastern part of Modoc County, California, and Mono Lake. I have one that I caught at Mono Lake in a sandy flat. While this is a widespread subspecies it does not seem to be very common anywhere. It seems to be an inhabitant of valleys and plains.

Perognathus parvus magruderensis Osgood. (Of Mount Magruder.)
MOUNT MAGRUDER POCKET-MOUSE.

Very similar to *olivaceous;* larger; skull larger and heavier; interparietal relatively narrower.

Length about 192 mm. (7.60 inches); tail vertebræ 102 (4); hind foot 24 (.95).

Type locality, Mount Magruder, Nevada, at 8,000 feet altitude; higher parts of the mountain ranges of the desert region of eastern California and Nevada, grading into *olivaceous* at their bases.

Perognathus alticola Rhoads. (A dweller on the heights.)
WHITE-EARED POCKET-MOUSE.

Above ochraceous buff thickly mixed with black; sides scarcely lighter than back; lateral buff line narrow and indistinct; black crescent at base of whiskers obsolete; ears and tail white; skull very similar to that of *olivaceous.*

Length about 165 mm. (6.50 inches); tail vertebræ 84 (3.30); hind foot 22 (.87).

Type locality, Squirrel Inn, San Bernardino Mts., California.

The half dozen known examples of the White-eared Pocket-Mouse have been taken in a small area in the mountains north of the town of San Bernardino, in open pine forest at 5,000 feet altitude or a little higher. They are not common or more would have been found with the amount of trapping that has been done

in the hope of getting more. They are easily recognized by
the ears and tail being white with but very little dusky markings.

Perognathus formosus Merriam.   (Comely.)
LONG-TAILED POCKET-MOUSE.

Size large; tail much longer than head and body; ears large;
the antitragus prominently lobed; above grizzled sepia; below
white; tail buff mixed with dusky above, buff below; cranium but
slightly arched; mastoids well developed; interparietal large and
wide; interorbital space wide.

Length about 190 mm. (7.50 inches); tail vertebræ 106
(4.15); hind foot 24 (.95).

Type locality, St. George, Utah.

Southwestern Utah, west to Owen Lake, California.

Perognathus longimembris Coues.   (Long membered.)
SAN JOAQUIN POCKET-MOUSE.

Above buff mixed with more or less black; below white; lateral line indistinct; tail buff, darker above, lighter below; skull
large; mastoids of moderate size; interorbital region narrow.
Young; darker and more olivaceous.

Length of male about 145 mm. (5.70 inches); tail vertebræ
74 (2.90); hind foot 19 (.75). Female smaller.

Type locality, old Fort Tejon, Kern County, California.

Southern part of the San Joaquin Valley.

Subgenus Chætodipus.

Mastoids relatively small and not projecting beyond the
plane of the occiput; pelage of adult harsh, often with spines or
bristles on the rump.

Perognathus penicillatus WOODHOUSE. (Pencil like.)
TUFT-TAILED POCKET-MOUSE.

Above pale clay color or vinaceous buff, sparsely intermixed with blackish hairs; below white; lateral line obsolete; hairs of terminal third of tail lengthened, forming a distinct "pencil" at tip; tail similar to back above, darkening toward tip, white below; soles naked; no spines on the rump, but sometimes small bristles are present; skull comparatively narrow; mastoids small; interparietal rather large, the angles rounded, transverse breadth nearly twice the longitudinal; interorbital region wide.

Length about 200 mm. (8 inches); tail vertebræ 109 (4.70); hind foot 25 (1).

Type locality, near San Francisco Mountain, Arizona.

This Pocket-Mouse is rare in the region where it was first found, that being in the edge of its habitat. It is more common along the Colorado River, but is not typical in the lower part of that region, gradually blending into the next subspecies.

Perognathus penicillatus angustirostris OSGOOD. (Narrow—rostrum.)
COLORADO DESERT POCKET-MOUSE.

Similar to *penicillatus;* averaging smaller; rostrum longer and more slender; color similar.

Type locality, Carrizo Creek, southwestern border of the Colorado Desert.

This Pocket-Mouse is common throughout the Colorado Desert, the southeastern part of the Mojave Desert, the southwestern corner of Arizona, northwestern Sonora and northeastern Lower California. The relative abundance is determined by the abundance or scarcity of plants and therefore of food. In considerable barren areas they are practically lacking, and in a few favorable localities they are abundant. I remember catching twenty-seven one night in a thick patch of weeds. Their food is mostly the small seeds of plants; mesquit beans are eaten to some extent. Their harvest is irregular and short, and the main dependance must be on stored seeds the greater part of the year. Much loss through starvation must occur after unfavorable seasons. The breeding season is spring, April to June. The usual number young is four and five, but I have taken several females containing seven fœtuses. In favorable seasons two litters of young appear to be raised.

Perognathus stephensi MERRIAM.
STEPHENS POCKET-MOUSE.

Similar to *penicillatus* but very much smaller; skull short; rostrum broader.

Type locality, the northwestern arm of Death Valley, California.

Known only from two examples which I caught in that part of Death Valley, known locally as Mesquit Valley. It is probably a dwarf subspecies of *penicillatus*.

Perognathus fallax MERRIAM. (Deceptive.)
SHORT-EARED POCKET-MOUSE.

Above brownish buff mixed with black; lateral line buff, usually well defined; white spot at base of ear small and faint; an indistinct dusky crescent at base of whiskers; feet and lower parts buffy white; tail pencillate and terminal third crested, upper side brownish becoming dusky terminally, white below; skull well arched; mastoids of moderate size; rostrum rather slender; interparietal large, wide; outer sides of nasals parallel to ends posteriorly; ears small, round; pelage coarse and mixed with long coarse spines on the rump and hips, those of the rump black and on the hips white.

Length about 185 mm. (7.30 inches); tail vertebræ 110 (4.33); hind foot 25 (1); ear from crown 7.5 (.30).

Type locality, Reche Canon, San Bernardino Valley, California.

Short-eared Pocket-Mice are common in southwestern California from the coast up to the lower edge of the pine belt. They occur some distance into Lower California and north to Los Angeles County. As they are nocturnal and seldom enter buildings, few people become acquainted with them, or know that these interesting animals are common about them. They principally inhabit weed patches and prefer sandy land. They rarely enter thick brush. Their tracks are often seen in dusty roads in the morning, and may be distinguished by the impressions of the long heels; frequently the mark of the tail may also be seen in the dust.

They live in burrows and under weeds and accumulations of dead leaves. The food is mostly seeds, with some leaves, buds and plant stems. Seeds are stored for food, commonly in small independent surface caches, not readily noticed. I have not heard them make any vocal sound, except rarely a squeak of pain. The number of young is usually five. These are born in April and May.

These Pocket-Mice make interesting pets. In November, 1889, I found one alive and unhurt in one of my traps and kept it a captive to study its habits. It was not wild but allowed me to handle it freely from the start. It would walk up my sleeve, around my neck and down the other arm, and for a year or more did not try to jump to the floor, but later it seemed to have lost the power to judge distances, and would jump down after a little walking about, even if the fall was great enough to injure it.

It never tried to bite me and would quietly bear handling and carrying about. I put it in a box with an inch or so of sand in the bottom; this it would scratch about vigorously in the night, but I rarely heard it moving in the daytime, although the interior of the box must have been fairly dark all the time. It did not try to gnaw the box, as true mice would have done, and did not try to lift the lid, which was kept closed by its own weight only. At first I tried feeding it grain, seeds and green food. It would eat no green plants or roots that I gave it and would not touch water. During the last three years of its life I gave it only dry barley or dry wheat and no water. It seemed to prefer the wheat. It is a mystery to me how such an animal can live for years and thrive on dry grain without water or moisture in any form, but this one certainly did. Three or four times a year I emptied the box and put in clean dry sand and set it in the corner of the hall, where it was perfectly dry, and put nothing more in the box but dry grain and a little cotton, of which the Pocket-Mouse made a globular nest.

If taken out of the box after dark and turned loose on the floor the Pocket-Mouse moved actively about a few minutes, usually by short, deliberate jumps; but if frightened it leaped two feet or more. After it had satisfied its curiosity it crept into a dark place behind some piece of furniture. If turned out on the floor in the daytime it hunted a dark place if allowed to, and was easily caught, but after dark I had to corner it to catch it. When captured this Pocket-Mouse appeared to be fully adult. It died in the summer of 1894, during my absence from home. It therefore, was at least five years old at the time of its death and probably older.

Perognathus fallax pallidus MEARNS. (Pale.)
PALLID POCKET-MOUSE.

Similar to *fallax* in size and proportions but paler, the intermixed dark hairs being fewer and brown instead of black.

Type locality, Mountain Spring, San Diego County, California.

The habitat of Pallid Pocket-Mice is the dry, cactus grown slopes of the mountains bordering the western side of the Colorado Desert, in San Diego County and northern Lower California. They do not seem to be common anywhere. They live among the rocks in the gulches and on the hillsides. They are associated with the Spiny Pocket-Mice in the lower part of the mountain slopes, but are easily distinguished from the latter species in the flesh. The animals found in the type locality of *fallax* are considerably paler than the usual run of *fallax,* and are really intermediate between normal *fallax* and *pallidus.*

Perognathus californicus MERRIAM.
CALIFORNIA POCKET-MOUSE.

Above yellowish bistre thickly mixed with black; spines of rump and hips prominent; lateral stripe pale fulvous, distinct; below yellowish white; tail long, crested-penicillate, sooty black above, white beneath; skull considerably arched; mastoids small; occiput bulging posteriorly; rostrum heavy; outer sides of nasals narrowed posteriorly; ears long, comparatively pointed.

Length about 193 mm. (7.60 inches); tail vertebræ 103 (4); hind foot 24 (.95).

Type locality, Berkeley, California.

Vicinity of San Francisco Bay, south to San Benito County.

Perognathus californicus dispar OSGOOD. (Unequal.)
ALLEN POCKET-MOUSE.

Similar to *californicus* but grayer and paler; grayer than normal *fallax;* skull heavier than typical *californicus* and in some cases with larger mastoids, thus approaching *femoralis.*

Length about 215 mm. (8.45 inches); tail vertebræ 120 (4.70); hind foot 27 (1.05); ear from crown 11 (.43).

Type locality, Carpenteria, Santa Barbara County, California.

Coast valleys of California from Los Angeles to San Benito County, and the western foothills of the Sierra Nevada north to Placer County. I have seen examples from the pine belt of the San Bernardino Mountains that were very large, with skulls that varied in the direction of *femoralis.*

Perognathus femoralis ALLEN. (Of the thigh.)
DARK POCKET-MOUSE.

Similar to *californicus;* darker and rather larger; skull similar to that of *californicus.*

Length about 220 mm. (8.65 inches); tail vertebræ 125 (4.90); hind foot 27 (1.05); ear from crown 11 (.43).

Type locality, Dulzura, San Diego County, California.

The Dark Pocket-Mouse has a somewhat limited range, as far as is now known. This is the foothills and mountains of San Diego County and the adjoining part of northern Lower California. I believe it will ultimately prove to be a subspecies of *californicus*. In the lower part of its range *Perognathus fallax* also occurs, and they may be trapped in the same spot. As they are very much alike the novice is likely to consider them both of the same species, but a little examination of the ears will show a difference; *fallax* has a short round ear while *femoralis* has a longer and pointed ear.

Perognathus spinatus MERRIAM. (Bearing spines.)
SPINY POCKET-MOUSE.

Above grayish buff mixed with dark brown; below white; lateral line obsolete; tail drab gray above, white below; spines large and extending forward on the sides, sometimes to the shoulders, the rump spines partly brown, the others white; skull rather flat; rostrum broad; mastoids small.

Length about 182 mm. (8.10 inches); tail vertebræ 110 (4.33); hind foot 21.5 (.85); ear from crown 6 (.24).

Type locality, 25 miles below Needles, California.

Spiny Pocket-Mice frequent the arid hills around the Colorado Desert and in the southeastern part of the Mojave Desert, east to the Colorado River, which they do not appear to cross. They are replaced on the east bank of the Colorado by the less spiny *intermedius*. Spiny Pocket-Mice are most plentiful along the western side of the Colorado Desert and southward into Lower California, but they are not plentiful anywhere. They do not seem to care for the open Desert or for wide valleys, but like the bottoms of rocky slopes where they can come into narrow sandy gulches. I have several times seen them running about among the rocks at twilight in summer, and they may be more crepuscular than other species.

Family **Zapodidæ.** Jumping Mice.

Skull of moderate size and thickness; occipetal region depressed; audital bullæ transverse, rather small; anteorbital foramen very large, oval, supplemented on the lower inner side by a small foramen, which transmits the second division of the fifth nerve; zygomatic arch depressed, the molar part slender, except anteriorly, where it widens and extends up on the maxillary to meet the lachrymal; upper incisors compressed, deeply grooved in front, orange colored; upper premolars present in one genus, absent in another; lower molars absent; molars rooted; enamel folds of grinding surface of molars complex; coronoid process of lower jaw high, slender, curved; angular process wide, twisted almost horizontal; cervical vertebræ not anchylosed; fore legs about half as long as hind legs; inner toe of front foot rudimentary; hind foot with five metatarsal bones and five toes, the inner toe short but functional; soles naked; tail slender, tapering, much longer than head and body; cheeks with internal pouches.

This small family is composed of two or three genera and about twenty species and subspecies. These inhabit the wooded parts of British America, northeastern Asia, Alaska and the northern parts of the United States, reaching some of the higher southwestern mountains.

Progress when hurried is by long leaps, sometimes seven or eight feet, but these long leaps soon tire the animal and the leaps shorten to a yard or less and the animal hides in the nearest cover. The long tail is a great help in making these long leaps and an aid in going in a straight line. Jumping-Mice do not make runways as many of the small animals are in the habit of doing. The food is vegetable, mostly seeds.

Hiberation occurs regularly in the greater part of the range of the family, but it is probably incomplete in the southern edge of their range. The animals become very fat, and the fall pelage is usually assumed before hibernation begins, which is when the first hard frosts occur.

Most species inhabit grassy valleys bordered by open forests or interspersed with groves or shrubs, and some prefer moist localities. They are most often noticed in mowing the grass of meadows inhabited by them. They are crepuscular and nocturnal, but are abroad occasionally in the daytime.

Genus **Zapus** COUES. (Great—foot.)

Nasals long, projecting some distance in advance of the incisors; upper premolar present, very small; enamel folds of molars crowded; frontal narrow interorbitally; ears rather long; pelage coarse; four pairs of mammæ.

Zapus trinotatus RHOADS. (Thrice—Marked.)
NORTHWEST JUMPING-MOUSE.

Summer pelage; a broad, well defined dorsal band from nose to tail black mixed with the color of the sides; head lighter; sides brownish ochraceous buff or yellowish clay color sparsely mixed with coarse black hairs, bordered below with a narrow buff line; feet white; tail bicolor, dusky above, whitish beneath. *Autumn pelage;* dorsal band more flecked with yellowish; sides dull yellow. *Immature;* back with less black.

Length about 240 mm. (9.45 inches); tail vertebræ 145 (5.70); hind foot 33 (1.30); ear from crown 11 (.43).

Type locality, Lulu Island, British Columbia.

Coast region of British Columbia, western Washington, western Oregon and northwestern California to Humboldt Bay.

Zapus trinotatus alleni ELLIOTT. (For J. A. Allen.)
ALLEN JUMPING-MOUSE.

Similar to *trinotatus;* dorsal band less black; skull smaller with small audital bullæ; tip of tail sometimes white. The autumnal pelage appears to be the same as that of summer. Size of *trinotatus.*

Type locality, Pyramid Peak, Eldorado County, California.

Mount Shasta and the Sierra Nevada. They frequent the mountain meadows and grassy localities along streams. They do not seem to be common in many places and I did not find them easily trapped. I got but four, three of these being taken in traps set for meadow-mice, and the fourth in a steel trap set in shallow water in a small stream below a spring in a mountain beaver runway. As the pan of the trap was close to or above the surface of the water it is probable that the Jumping-Mouse used it as a stepping stone in crossing the stream, which passed through coarse grass which nearly met over it. These four animals were taken at various altitudes from 5,000 to 9,000 feet. The breeding habits are probably similar to those of the eastern species, which sometimes have two litters annually, of four to six each.

Zapus orarius Osgood.
COAST JUMPING-MOUSE.

Sides of body and head dark ochraceous, moderately mixed with black; dorsal band not sharply defined and suffused with the color of the sides; lower parts strongly suffused with ochraceous, the sides of the throat deeper ochraceous; feet yellowish white; tail grayish above and yellowish white below; upper incisors slender and more projecting than usual; rostrum short and considerably deflected; nasals very narrow anteriorly; interorbital constriction narrow; audital bullæ small and rather near together.

Length about 220 mm. (8.65 inches); tail vertebræ 127 (5); hind foot 30 (1.20).

Type locality, Point Reyes, California.

Only known from the coast region of Caifornia from Point Reyes to Humboldt Bay. Evidently rare in this region.

Zapus pacificus Merriam. (Of the Pacific Coast.)
PACIFIC JUMPING-MOUSE.

Dorsal area not sharply defined, but so strongly suffused with yellowish that the yellow predominates over the black; sides buffy-yellow, moderately lined with black hairs; inner sides of legs only, slightly darkened; tail sharply bicolor, grayish above, white beneath; fore and hind feet soiled white.

Length of type 225 mm. (8.85 inches) tail vertebræ 141 (5.55); hind foot 31 (1.22).

Type locality, Prospect, Rogue River Valley, Oregon.

Southwestern Oregon, south to Mount Shasta, California. Rare.

Family **Erethizontidæ** American Porcupines.

Skull short, rugged, thick and strong; incisors large, prominent, not grooved in front; molars rooted and with complicated enamel folds; anteorbital foramen large, oval; angular process of lower jaw joining outside of root of lower incisor; coronoid process small and low; tibia and fibula distinct; tail prehensile in some genera, short and thick in others; toes variable in number with genera; soles tuberculate; pelage containing spines; upper lip not cleft; mammæ four.

The family contains about half a dozen genera and perhaps thirty species. But one genus and two species with several subspecies occur in North America. The food is vegetable, consisting chiefly of the fruits, twigs, leaves and bark of trees, shrubs and plants. Porcupines are plantigrade, principally nocturnal and more or less arboreal.

Genus **Erethizon** Cuvier. (To irritate.)

Four toes on front feet, five on hind feet, all armed with long, curved, compressed claws; tail rather short, very thick, not prehensile, covered above with stiff hairs and spines, and beneath with bristles; pelage below short and soft, above very long and mixed with sharp spines; spines hollow through the greater part of their length, small at base where they are loosely inserted in the skin, points sharp and over certain areas covered with minute flat scales pointing backward, thus acting as barbs.

Dental formula, I, 1—1; C, 0—0; P, 1—1; M, 3—3×2=20.

Erethizon epixanthus Brandt. (Outside—yellow.)
WESTERN PORCUPINE.

General effect yellowish gray, blackening on the rump, upper side of tail, face and feet; spines one to two inches long, very numerous, large creamy white with black tips; hairs sparse, four to six inches long, yellowish on the sides and on upper part of head and neck, whitish at base on the back with yellowish tip and black subterminal zone; face, fore legs and under parts blackish, free from spines; ears small.

Length about 760 mm. (30 inches); tail vertebræ 215 (8.50); hind foot 108 (4.25).

Type locality, northwestern North America.

Western Porcupines are found from the Sierra Nevadas to Alaska. They are common in a few localities ,but are almost unknown in large areas within their range. They occur south to the San Bernardino Mountains, one having been killed there in 1903, and they were occasionally seen there when these mountains were first known to white people. They seem to be lacking in all the coast mountains of Caifornia. They frequent the coniferous forests of the mountains and the higher valleys.

When wild fruits are in season these form the principal item of food of the Porcupines; such as wild gooseberries, wild currants, plums, etc. Cultivated fruits are also eaten, and I heard complaints of their damaging apple trees in the northeastern part of California by breaking off the smaller limbs in getting at the apples, and of the waste of apples, these having been dropped after having been bitten. When fruits are not obtainable these animals subsist on leaves, twigs, or sometimes on the inner bark of coniferous trees. They are fond of salt, and I saw a churn and a cheese press that had been gnawed by Porcupines, apparently to get a taste of salt. Townsend saw a pine stump eighteen inches in diameter that had been gnawed away by Porcupines; salt had been placed on the stump for horses to lick.

I found Western Porcupines common about Goose Lake, Modoc County, where they inhabited the crevices of lava cliffs. I saw one run from a thicket of wild gooseberries in the daytime and on shooting it found the stomach full of gooseberries. This one weighed nearly twenty five pounds. The gait is slow, one can easily outrun them.

Suborder **Duplicidentata**.

Three pairs of upper incisors, one pair being lost soon after the animals birth; enamel covering the sides as well as the front of the incisors; incisive foramina large and usually confluent; bony palate very narrow from front to back.

Family **Ochotonidæ**. Pikas.

Adult incisors two below and four above, the extra pair very small and hidden behind the others, which are deeply grooved in front; molars rootless; five toes on each front foot and four on hind foot; no visible tail; ears of moderate size, broad, rounded; eyes small; hind legs not greatly longer than the fore legs; size small.

This small family of one genus and fifteen or more living species inhabits alpine and boreal parts of Asia and North America. The species are all small in size and similar in habits. The food is grass and other herbage, the leaves and twigs of alpine shrubs, and probably some seeds. Food is stored in autumn for winter use, and it is probable that Pikas do not hibernate, although living where the snowfall is heavy. They are social; digitigrade; partly subterranean and principally diurnal. There is some variation in color, this being apparently due to environment and protective. The sexes are alike and the young but little different.

Genus **Ochotona** LINK. (Mongol name for the Pika.)

Occiput not depressed; a process of the molar is prolonged almost to the ear; aduitory bullæ large; angular process of lower jaw small, pointed, recurved; condylar process high, greatly flattened, wide antero-posteriorily; coronoid process minute, with a supplementary tubercle just back of the last molar; soles densely haired except a prominent pad at the base of each toe; pelage thick, long, coarse.

Dental formula, I. 2—1; C, 0—0; P, 2—2; M, 3—3×2=26.

Ochotona schisticeps Merriam. (Slate colored—head.)
SIERRA NEVADA PIKA.

Form thickset; no visible external tail, the short caudal series of vertebrae being folded back and lying wholly within the skin; ears prominent, broad, rounded, edged with white; whiskers very long; color of pelage variable with locality, from brownish gray mixed with black above and grayish vinaceous cinnamon below, to buffy white mixed with blackish on the back, the basal half of all the hairs being slaty black in both cases.

Sierra Nevada Pika. About half life size.

Length about 185 mm. (7.25 inches); tail vertebræ 13 (.50); hind foot 29 (1.15); ear from crown 17 (.67).

Type locality, Donner, Nevada County, California.

The habitat of the Sierra Nevada Pika is the higher mountains of California and northward. It has been found in various places in the Sierra Nevada, White Mountains, Mount Lassen, Mount Shasta and in the Warner Mountains. Probably it occurs in various other high mountains of California. Many years ago Prof. Gabb reported it from northern Lower California, but no one has found it there since.

I have taken Pikas in six localities in the Sierra Nevada and in the Warner Mountains. All these localities were similar and are typical of those the Sierra Nevada Pikas prefer. These are what are sometimes called "rock slides." On a steep slope where favorable rock occurs the frost and weather loosens blocks of rock, which roll down the mountain side, and the supply being continued the slope becomes covered many feet deep with blocks of rock from a few inches to several feet in diameter, generally assorted to size in certain parts of the slope. The angle of the slope is as steep as the blocks will lie. These slides often cover many acres. Among the interestices of these slides the Pikas make their homes, foraging on the herbage growing around the slides. They live in small communities, but my impression is that none of the localities that I have seen were inhabited by more than two or three dozen individuals, and in some probably but two or three families lived.

The food of Pikas is said to be "grass," but there is very little grass to be found in the neighborhood of any of the colonies that I have seen. Probably most of the plants growing within their reach are eaten. They are said to cut and cure grass for winter food; this "hay" being stored among the rocks after drying in the sunshine. The only instances of this kind that I have seen were on the Warner Mountains in July, when I found several piles of twigs, mostly of "Choke cherry" twigs. Some of these piles were of scarcely wilted twigs, while others were old, apparenly cut the previous season. These twig piles appeared so much like the nests of wood rats (*Neotoma*) that I got the impression that they were really used for shelter and secondarily for food when other food ran short. I have not had the opportunity of observing the Pikas late in the season when they would naturally be curing "hay."

The lowest altitude in which I have found Pikas is 6,700 feet and the highest about 10,000 feet. The Pikas run about on the rocks much as a rat would. I happened to see one on a bit of level ground at the foot of a slide; it hopped along much as a young rabbit would when not alarmed or hurried. The position taken at rest is that shown in our engraving. I have seen none standing upright. An adult that I weighed pulled the scale down to four ounces. Coues says of the Rocky Mountain Pikas that "reproduction takes place in May and June, and about four young are produced in a grassy nest." A female Sierra Nevada Pika that I shot August 22nd. was suckling young. The mammæ were six in number, two pairs pectoral, one pair inguineal.

The voice of Pikas is said to be similar to the bleat of a young lamb; that of the Sierra Nevada Pika is somewhat different. It may be represented by "eeh" strongly aspirated and is repeated. It is not loud but may be heard a hundred yards or more. The animal is difficult to locate by the sound, partly because of a ventriloqual quality of the sound, but more because of the animals resemblance to the rock on which it may be sitting. I have seen them through a field glass eighty or a hundred feet away when I could not distinguish them with the naked eye, yet a charge of fine shot fired at the top of the rock where I knew it sat would get the animal. The color is variable with locality, and protective, being similar to that of the rocks where they live. If the rocks are nearly white so is the animal, and if the rocks are dark the animal is also. The snowfall is heavy in the region inhabited by the Pikas, but the interstices of the rock slides would be free from snow and comparatively warm and it is probable that they are active all winter.

These little mountain dwellers are curious creatures from many points of view. They appear to be the remnant of an ancient family, distanced in the evolutionary race and crowded into a region difficult for other animals to utilize. They are peculiarly home loving bodies, as is shown by their having acquired the color of particular areas of rock, among which they must have dwelt many generations. Their habits protect them from nearly all predatory animals except weasels. American Pikas are sometimes called Little Chief Hares, Straved Rats and Conies.

Family **Leporidæ** Hares and Rabbits.

Skull long and narrow; two lower incisors; four upper incisors, the middle pair large and deeply grooved in front, the second pair small and hidden behind the first pair (in very young hares there is a minute third pair which are soon shed); molars rootless; rami of lower jaw wide and thin; facial surface of maxilla extensively perforated or reticulated; a perforation between the eye sockets; collar bones present but usually imperfect; size large.

The Hare family contains but two genera, about forty species and several subspecies. It is best represented in North America. There are several Old World species, but the family has no representative in Australia.

The food is entirely vegetable, mostly the leaves, stems or branches of a variety of plants and small shrubs, which are bitten off and eaten on the spot. In some regions they damage young orchards and vineyards, and occasionally they become sufficiently abundant to seriously harm grain crops. The family is of some economic importance as a source of food supply, the animals being of sufficient size to be worth hunting and the flesh being palateable.

Hares are digitigrade, terrestrial, principally crepuscular and nocturnal. In most species there is little change of color with age, sex or season, but some northern species undergo a complete change of color twice a year, being brownish in summer and white in winter.

The mammæ are numerous, usually five pairs, abdominal. The number of young are variable with species and region, the European Rabbit having several litters each year, each litter averaging half a dozen or more; while most American Hares have but two or three litters annually, and these seldom number more than four or five. The young of the European Rabbit are born blind and hairless in an underground nest, but the the American species of Hares are well haired at birth and can see and very soon care for themselves.

The name Rabbit properly belongs only to a particular species of Hare, the European Rabbit (*Lepus cuniculus*), which is well known in this country as the domesticated "English" Rabbit.

Genus **Lepus** LINNAEUS. (Hare.)

Hind legs longer than the fore legs; eyes large; ears long, usually equaling or exceeding the head in length; inner side of cheeks haired; tail short; soles heavily haired; no naked pads under the toes; pelage soft; skin thin; skull with distinct supraorbital processes; malar extending posteriorly in a short process; occiput depressed; auditory bullæ large; incisors very short, not reaching as far back as the premolars; coronoid process of lower jaw a thin, low, more or less incurved plate, sometimes obsolete.

Lepus campestris sierræ MERRIAM.
(Of the plains; of the mountains.)
SIERRA PRAIRIE HARE.

In summer; above grizzled gray, the hairs whittish for the basal two thirds, then blackish, then almost pure white and the tip again black; eye ring, front edge and part of inner surface of ear pale grayish buff; a broad stripe of the color of the head on the front side of the ear; back half of the convex side of the ear and nape white; tip of ear black; an indistinct white spot on fore head; breast gray; front side of fore legs and back side of hind legs pale buffy gray; soles brown; remainder of legs and belly white; tail large, bushy, white all around or with a narrow gray stripe on the upper side. *In winter;* white, more or less tinged with yellowish brown; ears tipped with black. Simetimes the change to winter pelage is incomplete.

Length about 135 mm. (25 inches); tail vertebræ 100 (4); hind foot 165 (6.50); ear from crown 150 (6); weight 6 to 10 pounds.

Type locality, Hope Valley, Alpine County, California, alt. 7800 feet.

Higher valleys of the Sierra Nevada south to Menache Meadows, in winter the eastern slope of the Sierras down to the upper edge of the sage brush. Dairymen summering in the Sierras told me that these Hares could not run fast. One man, who remained through the winters told me that they turned white with the early snows and showed me fragments of skins that were white on the surface and brownish or grayish beneath the surface. The large feet with unusually long hair on the soles are responsible for the peculiar name "Snowshoe Rabbits." They are also known as White tailed Jack-Rabbits. They do not seem to be plentiful anywhere.

Subgenus **Macrotolagus**.

Interparietal obliterated in adults; supraorbital process large, united to cranium posteriorly, in adults inclosing a large foramen; rostrum long; size large.

Lepus californicus GRAY.   (Of California.)
CALIFORNIA HARE—JACK-RABBIT.

Above grayish drab thickly mixed with black and tinged with fulvous; sides and breast grayish vinaceous cinnamon; belly buff or very pale cinnamon, ears drab, whitish on the back side, edged with brownish white or buffy white, and tipped with black; sides and under surface of tail grayish cinnamon, upper surface black, this stripe extending up on the rump; legs mostly light drab brown, more or less tinged with cinnamon.

Length about 560 mm. (22 inches) ; tail vertebræ 90 (3.50) ; hind foot 120 (4.75) ; ear from crown 160 (6.30) ; weight from four to seven pounds.

Type locality, probably the old San Antonio Mission, Monterey County, California.

Northwestern Lower California north through California

west of the Sierra Nevada, except the San Joaquin Valley, to southwestern Oregon. Common in the valleys and foothills of western California, occasionally ranging to the highest valleys of the coast mountains. They are not as gregarious as the Desert Hares, and are found less frequently in the open plains, preferring the edges of the plains and the little valleys in the foothills. In habits, food and gait they are similar to the Desert Hares. The breeding season is winter and spring. One New Years Day I shot a female California Hare that would have given birth to two young in about a fortnight. The number in a litter is two to four, and probably two litters a year are the usual number.

Lepus richardsoni BACHMAN. (For Sir John Richardson.)

RICHARDSON HARE.

Similar to *californicus* but smaller and much paler; above buffy gray.

Type locality, probably the Salinas Valley, Monterey County, California.

The Richardson Hare seems to be a pale species whose range is the Salinas Valley and the dry warm region eastward, bordering the San Joaquin Valley and extending north in the foothills of the Sierra Nevada to about Mariposa County. Its range overlaps that of the California Hare in the western part of the Salinas Valley.

Lepus texianus deserticola MEARNS. (Desert—dweller.)

DESERT HARE.

Above grizzled brownish gray; ears grayish brown on the front surface, fringed with grayish white or buffy white, tipped with black, remainder gray; hair on ears very short, sides of body, front side of fore legs and back side of hind legs pale gray, more or less tinged with buff or drab; belly whitish; upper surface of tail, extending more or less on the rump, black, remainder of tail pale gray.

Length about 510 mm. (20 inches) ; tail vertebræ 90 (3.55) ; hind foot 117 (4.60) ; ear from crown 155 (6.10).

Type locality, Colorado Desert.

Desert Hares are more or less common in the Deserts of southeastern California and northward along the eastern slope of the Sierra Nevada to the northeastern corner of this State.

Desert Hare.   One-fourth life size.

They are found in the open plains, the edges of the deserts and on the slopes at the foot of the mountains where small shrubs are scattered about, seldom in timbered places. Their food consists of nearly every kind of herbage obtainable, even cactuses being eaten, particularly in the drier parts of the year, when these are almost the only plants retaining any moisture. These Hares seldom drink, but obtain sufficient moisture to supply their bodily needs from the green plants eaten. Their run is graceful and very

rapid when at full speed. The gait when moving slowly is a ser-.
ies of hops and is ungainly.

The Desert Hares appear to be subject to epidemics, perhaps
more so than other species. In the summer of 1894, in Lassen
and Modoc Counties, I saw numbers of bodies scattered among
the sage brush and along the road. Some of those that I examined contained "warbles," but these were insufficient to account
for the death of the animals. It seemed to me that more died
than remained alive. Such epidemics have been frequently noticed, but I have seen no account of their occurence in central and
Southern California.

All Hares are subject to the attacks of numerous parasites,
such as tapeworms, ticks, bots and warbles. A brief mention of
some of these may be useful. In skinning a Hare a large blister
is sometimes found under the skin or in the flesh. This is sometimes called a "water blister." In the fluid contained in this blister are numerous larvæ of a tapeworm, a species of *Cænurus*. To
enable these tapeworm larvæ to complete the change to adult tapeworms it is necessary that they be transferred to the stomach of
some member of the *Canidæ*, as a dog or a coyote. This frequently occurs in the natural course of events. If you don't wish your
dog to suffer with tapeworm don't feed these blisters to it without
previous cooking. The very minute eggs of the tapeworm pass
from their canine host, and some adhering to vegetation are accidentally eaten by Hares, to continue this curious process of passing through two different animals to enable one parasite to complete its various life stages. The "warbles" spoken of above are
the larvæ of a species of *Cuterebra*, a fly which deposits its eggs
in the skin of the Hare, in the same manner that another species
does in the skin on the backs of cattle. Probably none of these
various parasites would render the flesh of a Hare harmful if the
fish is thoroughly cooked, but it is not appetizing to know of their
presence.

The young average about four, and it is probable that three
or more litters are born annually. Their fecundity must be great

to enable the species to hold its own, with birds and beasts prey-
ing on them, epidemics of disease decimating their ranks, and
man, with dog and gun, assisting to upset Natures balance. Fortunately for the Hare, man also destroys his hereditary enemy,
the coyote, else he would become exterminated.

Lepus texianus tularensis MERRIAM. (Of Tulare.)
TULARE HARE.

Very similar to *deserticola;* averaging paler and more yellowish; back less grizzled with black.

Type locality, Alila, Tulare County, California.

This is a pallid form inhabiting the southern part of the San
Joaquin Valley and the Carrizo Plain to the westward. It is
very abundant in many parts of this range and is the predominant
species captured in the large drives made in the San Joaquin Valley, in fact almost the only one taken in some of the drives. These
drives are large surrounds and the Hares are driven toward and
into corrals of wire netting. They are often very successful, several thousand being taken in a surround.

There are a few Hares in some of the higher mountains of
Southern California, in the more open parts of the pine forests.
They do not differ from *deserticola* sufficiently to be worth separa-
ting. They appear to be clearer gray and have shorter ears, judg-
ing from my scanty material, obtained in the San Bernardino
Mountains.

Subgenus **Sylvilagus**.

Interparietal present in adults as a small distinct bone; rostrum of medium length; skull and teeth light; suproarbital process
small, either united to cranium posteriorly enclosing a small foramen, or free; size medium or small.

Lepus auduboni BAIRD. (For John J. Audubon.)
AUDUBON WOOD-HARE.

Above clay color mixed with black, the tips of the hairs being black; sides brownish gray somewhat grizzled with black; inner (concave) surface of ears mostly pale gray, front part of convex surface brownish gray, tip black and remainder of outer side light gray, nape tawny ochraceous or pale iron rust color; an indefinite grayish white area around the eye; outside of legs light reddish brown, their inner sides white; soles light sepia brown; throat and belly white; underside of neck like the sides; tail comparatively long, its upper side similar to the back in color; on the hips the clay color of the back is replaced by dull white.

Length about 380 mm. (15 inches), tail vertebræ 50 (2); hind foot 88 (3.45); ear from crown 90 (3.55); weight about two pounds.

Type locality, San Francisco, also San Diego, California.

The Audubon Wood Hare is found from northwestern Lower California north along the coast region of California to some distance north of San Francisco and east to the Sierra Nevada and San Bernardino Mountains. It is common in much of this area and sometimes abundant. In eastern Califonia this form is replaced by the subspecies *arizonæ* and in northern California by *nuttalli*.

The Wood Hares are not often found in dense forests or in open plains. They prefer thickets of brush interspersed among trees, with some open ground about. They like to feed in the open ground, but hide in the brush when disturbed and in their hours of repose.

The food is preferably succulent herbs, but in places or seasons when these are scarce they eat most species of small plants, twigs and sometimes bark. Occasionally they do some damage to young orchards by biting off the branches or tops of small trees, rarely by gnawing large trees. They are frequently destructive to vegetables in gardens and grain crops. They can be prevented from damaging young trees by taking advantage of their dislike of the smell of blood and fresh flesh. The easiest way is to rub the body of the young trees with a piece of liver or freshly killed flesh, which they will avoid for weeks unless a heavy rain washes the trees.

The gait is a series of hops when moving about leisurely, and long rapid leaps when moving at full speed. Ordinarily they utter no sound, but when caught they make a harsh plaintive cry. The number of young in a litter is usually three to five; there are probably two or three litters annually, in spring and summer.

All our western Wood Hares take refuge more or less in crevices among rocks, but they do not often burrow in the ground. They rarely sit erect. They are timid creatures and depend on their excellent eyesight and hearing for warning and their speed or hiding in the brush for safety from pursuit. The Wood Hares are commonly known in the west as "Rabbits" and also as "Cottontails," which is a good general name for the group.

Lepus auduboni arizonæ. ALLEN. (Of Arizona.)
ARIZONA WOOD-HARE.

Paler than *auduboni,* the general effect gray rather than brown; ears longer and with but little black at tips.

Length about 355 mm. (14 inches); tail vertebræ 50 (2); hind foot 83 (3.25); ear from crown 92 (3.60).

Type locality, Beale Spring, in northwestern Arizona.

Western and southern Arizona, southern Nevada, California east of the Sierra Nevada and San Bernardino Mountains and northwestern Lower California. Frequents the thickets of the valleys and less arid parts of this region. Seldom common.

Lepus nuttalli BACHMAN.
NUTTALL WOOD-HARE.

Smaller than *auduboni;* ears shorter; color intermediate between *auduboni* and *arizonæ;* skull smaller; rostrum wider in proportion and much shorter.

Length about 335 mm. (13.20 inches); tail vertebræ 47 (1.85); hind foot 85 (3.35); ear from crown 80 (3.15).

Type locality, near junction of Snake and Columbia Rivers, Washington.

From northeastern California, eastern Oregon and eastern Washington east to the Rocky Mountains and the western part of the Great Plains. The Nuttall Wood Hare inhabits the sage brush region of northern California in Lassen, Modoc and Siskiyou Counties. It does not seem to be common in many places.

Lepus bachmani WATERHOUSE. (For John Bachman.)
BACHMAN BRUSH HARE.

Above grayish brown mixed with blackish, the back tinged with burnt umber; ears gray, darkest on outer surface, narrowly edged with whitish, rarely edged, but not tipped with black; nape light burnt umber; sides and throat brownish gray mixed with whitish; belly and front sides of legs pale gray, the plumbeous bases of the hairs showing through more or less on the belly; soles smoky brown; tail very small, its upper surface, sides and tip grayish brown, lower surface white; skull similar in size to that of *nuttalli;* condylar process of lower jaw shorter and more upright, angular process wider; compared with *auduboni* the same differences in these processes hold; the skull is smaller and the rostrum shorter.

Length about 330 mm. (13 inches); tail vertebræ 36 (1.40); hind foot 75 (2.90); ear from crown 67 (2.65).

Type locality, San Francisco or Monterey, California.

The Bachman Brush Hare is found in the coast region of California from Monterey north to Oregon.

· Lepus cinerascens ALLEN. (Ashy.)
ASHY BRUSH HARE.

Similar to *bachmani;* paler, the burnt umber tint of the upper parts nearly or quite absent; body smaller; hind feet shorter; tail and ears longer; teeth smaller; palatal bridge narrower; malar more depressed; condylar and angular processes of lower jaw narrower.

Length about 305 mm. (12 inches); tail vertebræ 36 (1.40); hind foot 72 (2.85); ear from crown 70 (2.75).

Type locality, San Fernando, Los Angeles County, California.

Southwestern California and northwestern Lower California. The Brush Hares inhabit thick brush, seldom venturing into open ground and rarely entering forests. The vast thickets of brush covering so much of the hillsides of California, and known locally as 'chapparal,' or more properly, 'chemisal,' forms their homes. They are very timid and are difficult to shoot because of their pertinacity in remaining in the shelter of the brush. In other respects their habits are similar to those of the Wood Hares.

In my notes I find records of two females each containing three fœtuses, March third and fourth respectively, and one with five, April seventh. Probably two litters are the rule. This Hare is well known to local hunters under the names of Brush Rabbit and Blue Rabbit, the latter name being given because of their bluish appearance at a little distance as compared with the Audubon Wood Hare.

Subgenus **Brachylagus**.

Skull short, deep; audital bullæ large; rostrum small; extremities of supraorbital processes free; ears, legs and tail short; size very small.

Lepus idahoensis MERRIAM.   (Of Idaho.)
IDAHO HARE.

Winter pelage; above clear drab gray slightly mixed with black hairs; ears pale buff inside, dull buffy ochraceous mixed with gray and black tipped hairs outside and bordered in front with a blackish line; nape and feet dull ochraceous buff; breast grayish buff; belly whitish along the middle line only. *Summer and immature pelage;* upper parts gray suffused with buff and intimately mixed with black.

Length 290 mm. (11.40 inches) ; tail vertebræ 15 (.60) ; hind foot 71 (2.80) ; ear 68 (2.28).

Type locality, Pahsimeroi Valley, Idaho.

The Idaho Hare is found from Idaho to northeastern California, a skin from Goose Lake being in the National Museum, and I saw a mounted specimen in Susanville which was probably of this species, but was unable to examine it closely. Dr. Merriam says that the Idaho Hare is strictly nocturnal and inhabits badger holes. It might therefore be common in a region and yet remain unknown.

Order **Feræ.** (The Flesh-eating Mammals.)

Digits never less than four to the foot, each bearing a claw; first and second digits not opposable as thumb and finger; teeth rooted, of three kinds, incisors, canines and molars; canine teeth prominent; condyles of lower jaw semiclindrical and placed transversely; clavicle incomplete or absent; radius and ulna distinct; scaphoid and lunar consolidated in one bone; stomach simple, pyriform; placenta deciduous and usually zonary; mammæ abdominal.

This order is often called the Carnivora because of the flesh eating propensities of its members. A few species live partly on fruits or vegetable materials, but all eat some flesh and the majority feed on freshly killed flesh almost exclusively. Many species rank high in intelligence and some are capable of useful domestication.

Suborder **Pinnipedia.** Seals.

Limbs short, fin-like, being useful only for swimming; first toe of front foot and first and fifth of hind foot longest; toes fully webbed; body prostrate; tail rudimentary; ears very small or lacking; eyes large and exposed, with flat cornea; no clavicles; skull constricted interorbitally; orbital fossæ very large; rostrum short and broad; milk teeth rudimentary and usually lost soon after birth; incisors varying in number with genus; canines lengthened.

Seals are abundant along Arctic seashores, common along those of temperate regions, but are less common in the tropics, where they are unknown on many coasts. The food of seals consists principally of fish, some species also adding crustaceans and mollusks. The fish are caught by pursuit in the water.

Seals are expert swimmers and spend the greater part of their time under water. They must come to the surface to breathe every few minutes, and they come ashore occasionally to rest or bask in the sun. Most species never venture more than a few

feet from the water. in which they quickly take refuge on be_
coming alarmed.

There are three families. two of these having species living
on the California coast, the third. *Odobenidæ*, containing only
one genus with two species, Walrus, being restricted to Arctic
seas. Altogether there are now about eighteen genera and thirty
species recognized in this suborder.

Family **Phocidæ**. Earless Seals.

Hind legs not capable of being turned forward and not ser-
viceable for use on land; front limbs smaller than hind limbs;
neck short; no external ears; upper incisors pointed; no distinct
posterbital process; uelage without underfur.

Subfamily **Phocinæ**

Incisors 3—2; all claws well developed; first and fifth toes
of hind foot not much longer than the other three; interorbital
region greatly constricted.

Genus **Phoca** LINN. (Seal.)

Molariform teeth, except the first, large, double rooted,
three lobed, planted more or less obliquely; head short; males
not much larger than the females; size small for the suborder.

Dental formula, I, 3—2; C, 1—1; P, 4—4; M, 1—1'×2=34.

Phoca richardii GRAY. (For Captain Richards.)
PACIFIC HARBOR SEAL.

Color variable; above yellowish gray, yellowish brown or
blackish, blotched with black, brown or buffy; below buffy whitish
or dull brown, more or less spotted with dark brown; skull thin
and comparatively smooth; premaxillaries extending to the nasals
and a short distance along them. The spots or blotches may be
very few and indistinct, or numerous and well marked, and are

commonly smaller and better defined on the lower surface, they may be lighter or darker than the ground color.

Seals appear to grow all their lives, but slowly after middle age. The length is three to five feet, rarely six. The females are a little smaller than the males, and the molariform teeth are somewhat smaller and less crowded.

Type locality, Vancouver Island, British Columbia.

Pacific Harbor Seals are common along the Pacific coast and islands of North America, from British America south probably to northern California.

Phoca richardii geronimensis ALLEN.
SAN GERONIMO HARBOR SEAL.

Similar to *richardii;* averaging larger; teeth heavier.

Type locality, San Geronimo Island, Lower California.

The San Geronimo Harbor Seals are common in many places along the California coast, particularly in the bays. They are monogamous. The young are born in May, June and July; one pup is the rule. They are not migratory and are not as gregarious as most seals are, being seldom found in large companies. They are comparatively silent, not making a loud roaring or barking as sea lions do.

Harbor Seals eat large quantities of fish, sometimes doing serious injury to inshore fisheries. In such places their killing should perhaps be encouraged, since the Harbor Seal is of but little use to civilized peoples, though to the northern Indians and Innuits they are an important source of food supply. The skin is of little value. The oil is not made use of here, though in some regions various small species are hunted for their oil. Scammon considers it the purest of all seal oils. Elliott says that the best seal flesh is that of the Harbor Seal.

These Seals are fond of basking in the sun, especially at low tide, when numbers may be seen lying on their favorite sand bars in the smaller bays that are not disturbed by shipping.

Progress on land, when hurried, is by pulling themselves.

forward by the fore feet used simultaneously, the hind feet being dragged along inactive. In swimming the hind feet ("flippers") do most of the work.

Subfamily **Cystophorinæ**

Incisors 2—1; first and fifth toes of hind foot longer than the other three and with rudimentary or no claws; interorbital region but moderately constricted.

Genus **Mirounga** GRAY. (Australian native name.)

Molariform teeth small, single rooted, not lobed; skull comparatively narrow; adult male with a nasal proboscis capable of voluntary elongation and dilation; webs of hind feet extending beyond the toes; adult males very large, the females much smaller.

Dental formula, I, 2—1; C, 1—1; P, 4—4; M, 1—1×2=30.

Mirounga angustirostris GILL. (Narrow—beak.)
CALIFORNIA ELEPHANT SEAL.

Light dull yellowish brown, varied with gray, darker on the back; more yellowish below; hind flippers hairy, without nails, deeply notched; foreflippers armed with long nails.

Length of adult males twelve to eighteen feet, the proboscis about fifteen inches in length; adult females are seven to ten feet in length and are without a proboscis.

Type locality, San Bartolome Bay, Lower California.

California Elephant Seals were formerly found along nearly the whole western coast of Lower California and north along the California coast to Point Reyes. It is probable that the species is now exterminated in the waters of this State, but a few may still survive about some of the outer islands of the Santa Barbara group. A very few are known to be still living on the Lower California coast.

California Elephant Seals seem to have frequented rocky

beaches in preference to sandy beaches or rocky islets. They are said to be monogamous. The young are born in May and June. Captain Scammon, to whom we are indebted for nearly all our information about this species, says that the sound made by them when alarmed resembled the lowing of an ox, but was more prolonged.

California Elephant Seal.

They were large and easily obtained, and produced a considerable amount of oil, two hundred gallons sometimes being obtained from a single individual. Therefore they were much sought after in the period succeeding the discovery of gold in California, previous to the general use of coal oil. In this period vessels were freighted with their oil, and in consequence of this reckless slaughter the species was practically exterminated in a few years. The California Elephant Seals while formerly abundant locally, were few compared with the enormous numbers of the Southern Elephant Seals that were found on islands of the Antarctic Ocean when these were discovered.

Family **Otariidæ** Sea Lions and Fur Seals.

Hind legs capable of being turned forward and of some service for use on land; neck long; front limbs nearly as large as hind limbs; first and fifth toes of hind feet without claws; webs extending beyond toes; small external ears; a postorbital process; incisors of upper jaw notched on crown; crowns of premolars and molars usually simple; males exceeding females in size.

Genus **Zalophus** GILL. (Great—crest.)

Molar not separated from premolars by a wide space; molariform teeth small, simple in root and crown; rostrum long and narrow; occipital and saggital crests prominent; very much so in old males; no underfur.

Dental formula, I, 3—2; C, 1—1; P, 4—4; M, 1—13×2=34.

Zalophus californianus LESSON.
CALIFORNIA SEA LION.

Color varying from tawny yellow through yellowish brown to dull black, the greater number being brownish yellow.

Length of adult males from seven to ten feet, and of females five to six feet; to end of outstretched flippers would be a foot more. I measured the large male spoken of later. His length to end of tail was 2490 mm. (98 inches); outstretched hind flippers 940 mm. (37 inches) from tip to tip. I estimated his weight at more than 500 pounds. I have seen larger seals of this species.

California Sea Lions are found along the Californian and Mexican coast from some distance north of San Francisco southeast to the Tres Marias Islands off the coast of southern Mexico. In 1884 I saw Sea Lions along the Sonoran coast of the Gulf of California which were probably of this species. These Sea Lions are not now found generally distributed along the coast of California, but resort to favorite rocky beaches on the Islands and to little islets along the coast where they are not molested.

These Sea Lions migrate but little, if indeed their movements may be given that name. At one season there seems to be
a dispersion from the breeding ground to the nearest well stocked
feeding ground, and at the reverse season a gathering at the
breeding ground. In some localities a good fishing ground is
found in the immediate vicinity of their favorite islets, which
they then occupy continuously. Like most of the larger seals this
species is polygamous. The sexes manifest but little attachment
for one another. The young are said to be born from May to
August, and at first are averse to the water, but in about a month
they enter the water and soon become expert swimmers.

The food is fish, which are swallowed without mastication. As
the Sea Lions are very expert and swift swimmers they are able
to overtake most fish by direct pursuit. But little economic use
is made of this Seal. At some seasons when they are fat some
are killed for their oil, and their hides are used for making glue.

The following extracts from my notebook may be of interest.
"April first, 1893. This morning Mr. Fenn and I rowed to the
'seal rocks' near the south end of Santa Catalina Island to get
sketches of the Sea Lions. As we neared the rocks we saw several Sea Lions on them and heard their loud 'hong-hong.'
Several were on the outer group of islets, but none were on the
outermost rock, which was perhaps a hundred feet from the islet
on which the greatest number were lying. The morning was cloudy
and calm, with but little sea, and we cautiously pulled up behind
the rock which rose five or six feet above the water. Mr. Fenn
got out on a little shelf that was awash when the larger swells
passed. In front the rock was low enough to see over and made
a rest for the sketch book. I had to keep clear of the rock to
avoid smashing the boat and out of sight as much as possible,
which was no easy job in the long swell.

"In another direction was a large rock about two hundred
yards away, on which a few Sea Lions crawled now and then,
but did not stay long; these I could watch as they were in full
view, though they did not appear to notice me at that distance.

They seemed to have hard work to crawl on the rock, and when up did not stay long in one position, but uneasily rolled about or slid back in the water. I saw one huge fellow crawl to a flat edge overhanging deep water and suddenly throw his fore parts up in the air and leap, just as a man would in diving, using his hind flippers to make the spring. Usually they slid down into the water head first.

"In the open space between the rocks and islets was a group of a dozen or so, rolling about in the waves and seeming to enjoy life hugely. They swam either side up or on their sides, heads rolling from side to side, flippers sticking out of the water here and there, a picture of perfect ease and contentment. Now and then they would tire of the idle pastime and commence romping; diving and chasing one another about, or leaping out of the water porpoise fashion. Once I saw a rather small one spring straight out of the water clear of the surface.

"At a distance a number of these Sea Lions crawling over the rocks, or slowly swaying their heads about as they often do, look like gigantic maggots, but at close quarters they lose that appearance. In the water the younger animals have a pleasant dog-like air as the head and shoulders appear above the water.

"The usual loud cry is a syllable sounding like 'hong' repeated sonorously, and has some resemblance to the hoarse bark of a dog. These sounds are oftenest made when the animal is lying on the rocks, but sometimes a swimmer takes up the bark. This occured several times near me. In these cases at least the animal did not close its mouth after each repetition of the note, but steadily held the mouth wide open, the head being held well out of the water. The vocalization all came from the throat, with no apparent modulation through action of the mouth or lips. Besides this repetition of the 'hong' I heard another sound a few times, a curious bleat, not loud, but to me, with a comical tone that always provoked a laugh. It always came suddenly, which was perhaps why it sounded so funny. This sound may have been made by a female.

"I wanted a skull of this species and when Mr. Fenn was through sketching we changed places. I selected the largest animal in sight, which fortunately laid on top of the islet in a flat place, and sent a rifle ball through the base of his skull, killing him so dead that he did not even struggle. I had feared that he would roll off and be lost. He was dark brown, except that a patch about three inches by four on the forehead and the region around the lips was a light yellowish gray. This light patch on the head seemed to be usual on all the light colored animals. The hairs of this patch are longer than those of the remainder of the head, and when dry project forward, giving a crested effect, lacking on the smaller animals.

"After satisfying our curiosity I cut off the head, which proved a tough job for my pocket knife, as the skin was nearly three quarters of an inch thick. We rolled the body off the rocks into deep water alongside, where it sunk as quickly as a stone and did not re-appear. While cutting the head off several medium sized animals came quite close to the islet as if they wanted to climb on.

"On our way back to camp we saw several small and medium sized Sea Lions and an old male playing in the surf. The male greeted us with barks and the whole group swam alongside us half a mile or more, as if to show us how agile they were, and how easily they could run away from us if they liked. Probably their motive was curiosity."

Genus **Eumetopias** Gill.   (Typical—broad forehead.)

Molar separated from premolars by a space about as broad as that occupied by a premolar; molar double rooted; premolars single rooted; crowns simple; rostrum short and wide; occipital and saggital crests not greatly developed, no underfur.

Dental formula, I, 3—2; C, 1—1; P, 4—4; M, 1—1'×2=34.

CALIFORNIA SEA LION

Eumetopias jubata SCHREBER.   (With a mane.)
STELLAR SEA LION.

Externally very similar to *Zalophus californianus;* supposed to average larger; colors similar and as variable.

Type locality, North Pacific Ocean.

Stellar Sea Lions range from Bering Straits east and south along the Pacific coast of North America to the Farallone Islands, California. I have never seen this species, but specimens of it from the Farallone Islands are in scientific collections. None are known from further south however, and this species is probably much less common there than the California Sea Lion. Externally the two species seem to be much alike, but probably an expert would see differences on comparison. The gap in the molars in the tooth row of the Stellar Sea Lion is quickly seen by any one who has an opportunity for such examination.

The habits of the two species are similar though the Stellar Sea Lion seems to migrate with more regularity. There is a difference in their voices, however, the utterance of the Stellar Sea Lion being a steady roar while that of the California Sea Lion is broken into barks.

Genus **Callorhinus** GRAY. (Beautiful—nose.)

Molariform teeth small, simple; rostrum very short and wide at base, convex in upper outline; saggital and occipital crests not greatly developed; pelage with abundant underfur.

Dental formula, I, 3—2; C, 1—1; P, 4—4; M, 1—1 or 2— 1'×2=34 or 36.

Callorhinus alascanus JORDAN and CLARK. (Bear like.)
NORTHERN FUR SEAL.

Male; black, the region over the shoulders gray; face with brownish areas; neck gray in front; flippers and belly reddish brown; *Female;* above gray; below rufous. *Young;* glossy black with more or less yellowish brown below.

Length of adult males about seven feet; female about four feet.

Type locality, Bering Sea, North Pacific Ocean.

The only present known breeding grounds of the Northern Fur Seal are the Pribilov or Fur Seal Islands of western Alaska, where Fur Seals are more or less abundant from June until November. The remainder of the year is spent in a great migration in the open sea, passing southward, eastward, northward and northwest to the breeding grounds again. Their great oval course brings them off the northern coast of California in early spring, though few are seen as far south as off the Farallones. When California was first settled by the whites Fur Seals were occasionally found on some of the islands and at a few localities on the mainland, but whether these were of this or the following species is not now known.

Genus **Arctocephalus** Cuvier. (Bear—head.)

Skull similar to that of *Zalophus* but with smaller occipital and saggital crests; molariform teeth six above and five below; pelage with abundant underfur.

Dental formula, I, 3—2; C, 1—1; P, 4—4; M, 2—1×2=36.

Arctocephalus townsendi Merriam. (For Chas. H. Townsend.)
GUADALOUPE FUR SEAL.

There are no skins of this species in any scientific collection and the species is definitely known only from the skull. There are four weatherworn skulls in the National Museum, only one being reasonably perfect. Dr. Merriam contrasts them with skulls of the Southern Fur Seal from the Galapagos Islands, finding the skull of *townsendi* to be smaller; rostrum shorter; nasals shorter, etc.

Type locality, Guadaloupe Island, Lower California.

The Guadaloupe Fur Seals are known to still occur in very small numbers at the Guadaloupe Islands. It is believed that they formerly occured on the Santa Barbara Islands, and there is a bare possibility that they may be found there again.

Suborder **Fissipedia**. Terrestrial Carnivores.

Limbs long, mobile, adapted for walking; toes free, with
long, sharp claws; first and fifth toes not longer than the others:
external ears well developed; incisors three above and three below,
on each side of the middle; a highly specialized premolar or molar
('sectorial') cutting tooth behind the middle of each jaw.

Family **Felidæ**.   (Cats.)

Digitigrade; front feet with five toes; hind feet with four
toes; claws retractile; tail long or short; upper surface of tongue
covered with sharp points, rasp-like; skull very short and broad;
rostrum very short; teeth 28 or 30.

The family of Cats is very generally distributed over the
world except in Australia. There are three living genera, and
over fifty species are now known. The food is principally the
flesh of animals caught by themselves.

Cats are principally nocturnal in habit, terrestrial or somewhat arboreal. The seasonal changes of pelage are not great, the
sexes are alike in some species, more or less unlike in others, the
young often differ in color from adults and are usually spotted.

Genus **Felis** Linn.   (Cat.)

Tail from one fourth to one half the total length; ears not
tufted; legs of moderate length; upper sectorial tooth (last premolar) very large, with three cusps and an inner tubercle on a
separate root; upper molar very small and placed inside the back
corner of the sectorial tooth; front upper premolar very small;
incisors small; brain case narrow; temporal crests parallel and
nearly united along the saggital suture; postorbital processes
small.

Dental formula, I, 3—3; C, 1—1; P, 3—2; M, 1—1×2=30.

Felis hippolestes olympus Merriam.   (Horse—robber, of the Olympic Mountains.)

PACIFIC COAST COUGAR.

Above varying from rufous brown to grayish tawny according to season, locality and probably individually (dimorphism), darkest on the back and tail; face dusky brown; a pale spot over each eye; convex side of ears blackish except the back side, which is gray; a blackish patch at base of whiskers; lips and chin white; neck dull fulvous, palest below; breast and inside of thighs dirty white; end of tail blackish. *Newly born young;* back and legs spotted and tail ringed with dark brown.

Length about 2135 mm. (84 inches); tail vertebræ 700 (28); hind foot 250 (10).

Type locality, Olympic Mountains, Washington.

The Pacific Coast Cougar ranges over the Pacific coast region from British Columbia to northern California. Cougars are not often found in dense forests and seldom in open plains, but prefer hilly localities with some timber and brush for cover. In some few localities they are sufficiently common to be troublesome to stock, but in the greater part of the west they are so rare that comparatively few people can say that they have seen a Cougar alive at large. I have knocked about in the wildest parts of the west much of the last thirty years and I have yet to see a live Cougar outside of a cage. I have seen their tracks in sand or dust a few times in out-of-the-way places and that is all. I once heard a distant cry that I suppose was made by a Cougar. It was in the night and was a loud wailing cry that in the distance had an unpleasant human tone.

The food of Cougars is flesh exclusively, which they prefer to kill themselves, rarely eating carrion unless forced by hunger. Being strong and powerful they prefer large game, such as deer, colts, hogs, grouse and turkeys. They are not ferocious and kill only for food. Instances of their attacking people have occurred, but these are very rare and they have learned to avoid man. In South America there is a widespread belief among the Indians that the Cougar will protect man from other animals, such as the Jaguar. Cougars have frequently been tamed, and have become interesting pets when not teased.

The flesh of Cougars is said to look and taste like veal. The number of young is commonly two, but sometimes three and four. They have bred in captivity several times. The Cougar is known by a great variety of names. Among these are Panther, Painter, Red Tiger, Puma, Catamount, American Lion, California Lion, Mountain Lion, etc. The Indian name Puma is perhaps the best one and Cougar next.

Felis aztecus browni MERRIAM. (For Herbert Brown.)
BROWN COUGAR.

Paler and grayer than *olympus;* teeth and audital bullæ smaller; size probably averaging a little less . A young Cougar that I saw in a cage was tawny drab gray above; lower side of head, breast and back part of belly whitish; tail like back except dusky tip; a blackish spot at base of whiskers; nose to eyes brown; a short perpendicular black stripe over each eye.

It was about 45 inches long; tail 15; hind foot 7; ear a little less than 2. It was probably about four months old.

Type locality, near Yuma, Arizona.

Brown Pumas range over northern Lower California, southern California, Arizona and the Colorado Valley. They are scattered through the mountains and in the timber of the river bottom in the Colorado Valley.

Genus Lynx KERR. (Sharp sight.)

Legs long and strong; tail very short; ears usually tufted; a ruff of long hairs on the neck; but two pairs of upper premolars; teeth otherwise similar to those of *Felis;* brain case broad; temporal crests widely separated; postorbital processes long, almost meeting; pterygolds long and very slender.

Dental formula. I, 3—3; C, 1—1; P, 2—2; M, 1—1×2=28.

Lynx eremicus Mearns. (Hermit.)

DESERT LYNX.

Summer pelage (May to Sept.) ; above grayish tawny olive more or less mottled or spotted with brown or blackish, usually with a pair of narrow interrupted black stripes along the back; crown with indistinct narrow blackish stripes; an indefinite whitish eyering surrounding black eyelids; whiskers mostly white, with several rows of small black spots or lines at their basis; convex surface of ears black enclosing a triangular pale gray spot, which often covers half the ear; upper side of the tail similar to the back, with black tip and one to six black bars, the intervening spaces becoming whitish toward the tip; under side of tail white; outside of the legs more tawny than the back, indistinctly spotted with brown or blackish; inner side of legs grayish or brownish white distinctly barred and spotted with black; throat white sometimes spotted with blackish; a wide buffy or tawny band across the breast, usually spotted with blackish; belly white spotted with black. *Winter pelage.* (Sept. to May) ; tawny shades of upper parts replaced with drab gray. *Young;* at first tawny thickly spotted with blockish, the spots small; later the color becomes grayer and the spots larger and fewer.

Length about 825 mm. (32.50 inches) ; tail vertebræ 160 (6.30) ; hind foot 170(6.70) ; ear from crown 80(3.15) ; the tuft of hairs nearly an inch longer. Weight 12 to 20 pounds. The female averages smaller than the male.

Type locality, Colorado Desert, California.

Desert Lynxes are common in most of the wooded and brushy parts of central and southern California from the seacoast to the Colorado River, and in northern Lower California. They vary greatly in amount of spotting, shade of color, size of ear tuft and barring of tail, dependent on age, season and wear of pelage. Before me lie seventeen Lynx skins taken in one locality (35 miles northeast of San Diego), all prepared and measured by myself, therefore strictly comparable. These vary in color from a tawny olive above with scarcely an indication of spots, to drab gray

thickly spotted and barred with black everywhere, with intermediate examples fully connecting one extreme with the other. The
tails of two have black tips with no bars, while others have two
to six bars. In addition to these are two Colorado Desert examples which I am unable to separate subspecifically from the
coast form. I have seen a number of others from the Desert also.

Desert Lynxes prey on all the smaller animals and birds and
frequently on poultry. They ordinarily prowl about at a walk. If

Desert Lynx.

pressed they can run rapidly a short distance, but if chased by
dogs they soon run up a tree. They can leap a considerable distance.

The only vocal sounds that I have heard from Lynxes are
threatening growls if approached while fast in a trap, at which
time they appear quite formidable. They 'spit' like a domestic
cat. The senses are all acute. They do not appear to be very
courageous. I can learn of but three instances of their having at-
tacked people, all occuring in San Diego County. Two of these

attacks were unprovoked and the third was practically in self defence.

Lynxes do the greater part of their hunting at night, yet they prowl around more or less in the daytime. Much of the poultry they get is caught in the middle of the day. I have trapped successfully for Lynxes by putting the bait in the back part of a crevice of rock, between roots of trees or in V-shaped openings of thick brush, placing the traps two or three feet in front of the bait. Loose rocks or brush are placed at the sides of the traps to guide the cat over them. I usually put two traps in front of each bait, one in front of the other. An old hen is good bait, but any fresh flesh will answer. The bait should be renewed daily. This is a good method of trapping for all carinivores.

The young appear to be born in March and April. About the end of May I saw two kittens playing about the crevices of a rocky cliff and trapped one, shooting the parent as she was lying on a rock above the cliff. The kitten was about a month old and very little injured so I kept it alive some time. It did not become reconciled to captivity and would jump against the wire netting with a spiteful growl and spit every time I came near.

Lynx fasciatus pallescens MERRIAM. (Banded; pallid.)
WASHINGTON LYNX.

"Similar to *fasciatus* but slightly smaller and much paler; general color hoary gray, constrasted with the dark rich rufous of *fasciatus*. Larger than *eremicus* and with much larger teeth.

Type locality, Trout Lake, Washington.

I have not seen this Lynx. Dr. Merriam states that it is common around the base of Mount Shasta.

The Canada Lynx is not found in California, though hunters frequently report killing them. Many people think that every Lynx with a tuft of hairs on the ears is a Canada Lynx. The animals of the genus *Lynx* are often called Wildcats. This name is best restricted to the long tailed *Felidæ*.

Family **Canidæ** (Dogs, Wolves, Foxes.)

Digitigrade; front feet with five toes, the inner toe small and placed some distance above the others; hind feet with four toes; claws not retractile; tail long, usually bushy; tongue normal; rostrum very long; palate not extending much back of molars; teeth usually 42.

This is a large family of nearly a dozen genera, probably seventy-five living and several extinct species. The family is very generally distributed over the world, including Australia. All are carnivorous, some species exclusively so, but many species eat other food such as fruits.

Genus **Canis** LINN. (Dog.)

Pupil of eye circular; tail moderately bushy; upper incisors with a distinct lobe each side of the main point; incisors of moderate size; front cusp of upper sectorial tooth obsolete; first upper molar very large and the second one rather large, each with two prominent outer cusps and a lower tubercular shelf supported on an inner root; lower sectorial tooth (first molar) very large; last lower molar and first lower premolar very small; temporal crests united in one saggital crest; postorbital processes small; angular process of lower jaw long, curved, distinct.

Dental formula, I, 3—3; C, 1—1; P, 4—4; M, 2—3'×2=42.

Canis ochropus ESCHSCHOLTZ. (Ochre—foot.)
VALLEY COYOTE.

Above buffy ochraceous mixed with black; below whitish more or less tinged with buff and with the long hairs of throat and breast more or less tipped with black; ears whitish on the inside and tawny or tawny ochraceous on convex surface; nose grayish cinnamon; legs dull tawny ochraceous, paler or whitish on the inside and grizzled with black tipped hairs on the outside next the body; tail, above like the body, below dull tawny ochraceous whitish at base and darkening toward the tip which is black, a black spot on the upper side about midway of its length; skull narrow, the nose slender; teeth small for the genus, with wide spaces between the premolars.

Length about 1133 mm. (45 inches); tail vertebræ 300 (12); hind foot 180 (7.10). Female smaller.

Type locality, California, probably in the central part of the State.

Valley Coyotes inhabit the region between the Sierra Nevada and the Pacific Ocean. This race probably does not go high in the mountains, but frequents the foothills and valleys, where they are common. As Coyotes obtain a considerable part of their prey by running it down they hunt principally in open ground, where their game has less opportunity to evade them. Coyotes eat almost anything in the way of flesh obtainable, including carrion though they prefer fresh meat. They eat beetles, grasshoppers and other insects, grapes and other fruit. They are often a serious pest to sheep owners as mutton is a favorite article of diet with them.

On the other hand Coyotes are a material help to the vineyardist by keeping hares in check, though they sometimes help themselves to a few grapes or rarely a chicken, in payment of their services. It will be remembered that a few years since the State Legislature passed a law giving a bounty for the destruction of Coyotes, with disastrous results to the State treasury and no lasting benefits to the sheep and cattle industries. One of the results of this unwise law was the great increase of hares and the giving of bounties by several Counties for the destruction of hares.

Coyotes are swift of foot and able to maintain a rapid run a considerable time. They show considerable skill and cunning in hunting tobether, but not to the extent that is sometimes credited to them. They rarely hunt in large packs, seldom more than three or four in company, and most frequently alone.

Their voice is similar to that of a dog, for which reason they have been called Barking Wolves. Their usual series of notes is several short barks followed by a longer note that may be termed a short howl, but no one familiar with the howl of the gray wolf would call the Coyotes song a howl. To me the barking of a Coyote at a little distance is a pleasant sound, probably because of its association with the memory of many a camping trip in plain and desert. Coyotes never attack human beings, and one need be no more afraid of them than of a hare or a fox. Baird says that Coyotes sometimes have as many as ten young in a litter and that they are born in April. I have no other data at hand, but my impression is that the number is generally much smaller and that they are born in March and May also.

Canis estor MERRIAM. (Eater.)
DESERT COYOTE.

Similar to *ochropus* but paler; buffy, gray above grizzled with black; skull shorter and wider and teeth rather heavier.

Length about 1060 mm. (42 inches); tail vertebræ 300 (12); hind foot 180 (7.10); ear from crown 125 (5). Female smaller.

Type locality, San Juan River, southeastern Utah.

Desert Coyotes are common from the eastern slope of the Sierra Nevada and San Bernardino Mountains east to the southern Rocky Mountains, and in southern California. Their habits are similar to those of the preceding species.

Canis lestes MERRIAM. (Robber.)
MOUNTAIN COYOTE.

Similar to *ochropus* but larger and somewhat paler; ears smaller; tail broadly tipped with black; skull larger and heavier with nose much broader; teeth larger.

Length about 1170 mm. (46 inches); tail vertebræ 320 (12.60); hind foot 200 (7.85). Female smaller.

Type locality, Toyabe Mountains, central Nevada.

Mountain Coyotes are found in the higher mountains of northern California, in the Sierra Nevada and in the mountains eastward to the Rocky Mountains. In the interior northward they are found to the plains of British America. In California in summer this species seems to be limited to the mountains and high plains of the northeastern part of the State, coming lower in winter.

Canis mearnsi MERRIAM. (For Dr. E. A. Mearns.)
MEARNS COYOTE.

Similar to *ochropus* but deeper colored, more tawny; skull and teeth similar to those of *estor;* size of *estor*.

Type locality, Pima County, Arizona.

Southern Arizona, northern Sonora, the mountains of northern Lower California and of southern California.

This brightly colored coyote occurs in small numbers in the mountains of San Diego County.

Little harm would be done if all the Californian Coyotes were grouped together under the name *ochropus*. There is considerable individual and seasonal variation of color, and the skulls, which are the best guides, in most cases require careful comparisons with good series to decide which species the individuals belong to. Those not experts will find it difficult to determine the species of Coyotes.

Canis mexicanus LINN.
GRAY WOLF.

Size large; general color varying from very pale gray to blackish; commonly yellowish gray above, more or less grizzled with long black hairs, the amount of black being quite variable; lower parts pale yellowish gray, sometimes with black tips to the hairs of the neck; ears brownish yellow or reddish; tip of tail blackish.

Length about 1525 mm. (60 inches); tail vertebræ 380 (15); hind foot 230 (9).

Type locality, Mexico.

A very few Gray Wolves live in the high Sierras and in the mountains of northeastern California. I do not know of any California example in any museum or private collection.

Genus Vulpes FINSCH. (Fox.)

Pupils of eye elliptical; tail long and bushy; upper incisors rather small, not distinctly lobed; teeth otherwise similar to those of *Canis;* temporal crests low, parallel, not widely separated; upper postorbital process small, the corresponding process on the zygomatic arch obsolete; angular process of lower jaw short, narrow, curved.

Dental formula, I, 3—3; C, 1—1; P, 4—4; M, 2—3×2=42.

Vulpes macrotis MERRIAM. (Great—ear.)
LONG-EARED FOX.

Small; ears large, color very pale, above pale grayish buff; chest and fore legs buff; remainder of lower parts buffy white; tip of tail and a small spot on upper side near base chestnut or sepia.

Length about 760 mm. (30 inches); tail vertebræ 290 (11.40); hind foot 120 (4.75); ear from crown 95 (3.75). Weight four pounds.

Type locality, Riverside, California.

Long-eared Foxes are a desert species. The type came from the western edge of their range. They are not very common in any part of their range. They are found in southwestern Arizona, in the Colorado and Mohave Deserts and a few straggle through the San Gorgonio Pass west and south. They live in open, nearly level localities, quite the reverse of the habit of our other Foxes. They live in burrows, these often having several entrances. They are nocturnal and crepuscular, but are often abroad some time after sunrise. They are not hard to trap. I do not remember hearing any bark or other vocal sound that I could attribute to them. Their food seems to be the small rodents living in the region that they inhabit. I have caught young of this species about the middle of April that were about a month old. There were three in this litter and this is probably about the usual number.

Vulpes muticus MERRIAM. (Unarmed.)
SAN JOAQUIN FOX.

Similar to *macrotis* but larger; hind foot and tail longer; back browner; skull larger; rostrum much broader; teeth larger.

Length about 950 mm. (37.50); tail vertebræ 350 (14); hind foot 125 (5).

Type locality, Tracy, San Joaquin Valley, California.

So far as I know this Fox is found only in the San Joaquin Valley.

Vulpes necator MERRIAM.
HIGH SIERRA FOX.

Red pelage; face dull fulvous, strongly grizzled with whitish; sides of nose dusky, grizzled with buffy; upper parts from back of head to base of tail dark dull dusty fulvous, becoming much paler on the sides where the whitish underfur shows through; black of fore feet reaching up on the upper surface of fore leg to elbow; black of hind feet ending at or near tarsal joint, with only slight traces on the outer side of leg; tail at base fulvous, becoming buffy whitish and profusely mixed with long black hairs; base with the usual black spot; tip white. *Black-cross pelage;* back grizzled black and whitish or buffy; sides buffy; feet, legs aid belly black; tail mainly black with white tip.

Length about 965 mm. (38 inches); tail vertebræ 370 (14.50); hind foot 160 (6.30).

Type locality, Whitney Meadows, Sierra Nevada, California.

Known only from the high Sierra from 6,000 feet altitude upward.

Vulpes cascadensis MERRIAM.
CASCADE MOUNTAIN FOX.

Red pelage; general color of head and upper parts straw yellow; face from nose to eyes dull yellowish fulvous; rest of top of head and base of ears pale straw yellow; back golden yellowish fulvous; tail very pale; black of ears restricted, that of the feet confined to the upper surface and mixed with pale fulvous. *Black-cross pelage;* top of nose grizzled brownish; sides of nose and imperfect ring around the eye dusky or blackish grizzled with whitish; top of head yellowish white, the black underfur showing through; dorsal cross (back and shoulders) blackish, overlaid and nearly concealed by yellowish white or buffy; sides of neck, flanks, and post-scapular region golden yellow; upper two thirds of ear black; fore feet black, grizzled above the elbow with yellowish; hind feet and legs grizzled dusky and buffy, becoming nearly black on top of the feet; chin, throat and band down middle of belly black or blackish; tail black mixed with buffy and tipped with white; skull as compared with that of *necator* has a wider rostrum, the audital bullæ are larger and the teeth are smaller.

Length about 1070 mm. (42 inches); tail vertebræ 410 (17); hind foot 173 (7).

Type locality, Trout Lake, Cascade Mountains, Washington.

The Cascade Mountain Fox is found in the Cascade Mountains of Washington and Oregon and in the northern Sierra Nevada. This species and the High Sierra Fox belong to the group which in one pelage is called the Silver Gray Fox and in another pelage the Cross Fox. The latter name is due to the fact that a dark stripe on the top of the neck and back is crossed by another on the shoulders. The Silver Gray Fox is black with the hairs, particularly on the hips, with long white tips, producing a silvered appearance. This stage of pelage is probably very rare in California, if it really occurs at all. These different pelages are not seasonal but are to a considerable extent dimorphic, though there is also some seasonal differences.

Genus **Urocyon** BAIRD. (Tail—dog.)

Pupils of eye elliptical; tail long, bushy, with a concealed mane of stiff black hairs on its upper side; teeth as in *Vulpes;*

temporal crests distinct, widely separated, converging posteriorly; upper postorbital process large and the corresponding process on the zygomatic arch nearly as large; lower jaw but slightly curved; the angular process long, wide, curved, and with a deep wide notch below anteriorly.

Dental formula, I, 3—3; C, 1—1; P, 4—4; M, 2—3$\times$2=42.

Urocyon californicus MEARNS.
CALIFORNIA GRAY FOX.

General effect above grizzled gray, underfur grayish buff, darkest on the head, lightest on the sides, thickly mixed with long hairs which are black in the basal two thirds covered by the under-

California Gray Fox.

fur, then white, then black to tip, the black tip being shorter and paler on the sides; upper part of head similar to the back; upper side of nose nearly to eyes mixed brown and gray; sometimes quite pale; a white area in front of whiskers, this usually including the point of the under lip; remainder of lower jaw black, this area joining the black area at the base of the whiskers; inner surface and edges of ears whitish; convex surface of ears mixed dusky, gray and tawny near the tip changing to clear tawny, ochraceous or brownish buff toward the base of the ear, this color extending on the sides of the neck nearly to the shoulder, outer and back sides of legs tawny or ochraceous; remainder of under side of head and neck, inside of legs and more or less of breast and belly white, front side of fore legs and feet and of hind feet mixed white and dusky; tail with a stripe of long stiff black hairs on the upper side, a fulvous stripe on the underside, the remainder like the back.

Length about 950 mm. (37 inches); tail vertebræ 380 (15); hind foot 130 (5.10); ear from crown 82 (3.25). Weight 8 pounds.

Type locality, San Jacinto Mountains. California.

California Gray Foxes are found in nearly all the forested parts of central and southern California, but are not common in many places. I found them up to 9,000 feet altitude in the San Jacinto Mountains. Their food is small mammals, birds, insects and fruit. I have heard but few complaints of these Foxes destroying poultry. They are not very difficult to trap. Their bark is hoarse and not loud. The young are born in April, May and June.

Urocyon californicus townsendi MERRIAM. (For C. H. Townsend.)
TOWNSEND GRAY FOX.

Similar to *californicus;* ears larger; tawny parts deeper colored; rostrum broader; teeth heavier.

Type locality, Baird, Shasta County, California.

Common around Mount Shasta and probably in most suitable places in northern California.

Urocyon littoralis BAIRD. (Of the seashore.)
SAN MIGUEL ISLAND FOX.

Very similar in color to *californicus* but size very much smaller.

Length about 725 mm. (28.50 inches); tail vertebræ 250 (10); hind foot 110 (4.33). Weight about four and a half pounds.

Type locality, San Miguel Island, Santa Barbara group, California.

Urocyon littoralis santacruzæ MERRIAM.
SANTA CRUZ ISLAND FOX.

Similar to *littoralis* but colors brighter, skull smaller; nasals narrower; rostrum narrower.

Length about 710 mm. (27.90 inches); tail vertebræ 265 (10.35); hind foot 108 (4.25).

Type locality, Santa Cruz Island, California.

Urocyon clementæ MERRIAM.
SAN CLEMENTE ISLAND FOX.

Similar to *littoralis* but skull smaller; nasals more tapering posteriorly; rostrum more slender.

Type locality, San Clemente Island, California.

¹ Urocyon catalinæ
SANTA CATALINA ISLAND FOX.

Similar to *littoralis* but tail longer; nasals narrower and not constricted in the middle; rostrum longer.

Length about 760 mm. (30 inches); tail vertebræ 285 (11.25), hind foot 112 (4.40); ear from crown 63 (2.50).

These various forms of Island Foxes are so much smaller than our mainland Foxes that anyone can distinguish between them at sight, but when it comes to discriminating the various island forms from one another it is a different matter, and most people will name them from the island they are found on without regard to technical characters.

Island Foxes are or have been common on nearly all of the

Santa Barbara Islands, and on some they were abundant. They seem to be subject to an epidemic of some kind. In 1886 I found them abundant on Santa Catalina Island. In 1893 I spent nearly a month at the same place and saw none and was told by residents that they were rare but appeared to be increasing. In 1886 I saw two and three together several times. They were not shy and moved about in the daytime to some extent. Several times I heard a bark in the night that sounded much like that of the gray fox of the mainland. Those foxes that I skinned had many cactus thorns in their skin and flesh. The flat leaved cactuses (*Opuntia*) are very abundant on the island and of necessity the Foxes get pricked in pursuing squirrels and birds in the cactus thickets, to which these resort for protection. The Foxes patrol the beaches to pick up any fish that may wash ashore. I saw no burrows and suppose that the Foxes spend the day in the dense thickets of brush on the hillsides.

Family **Procyonidæ.** (Raccoons, etc.)

Plantigrade or digitigrade; five toes on all the feet; tail long, bushy, usually annulated with rings of different colors; rostrum moderately long; molars tuberculate; teeth 36 to 40.

This is a small family of five or six genera and about a dozen species. One genus (*Ælurus*) of southern Asia is placed in a subfamily by itself and may not belong to this family; all the other genera are limited to temperate and tropical America.

They are principally nocturnal, arboreal and terrestrial. They are but partly carnivorous, eating small birds, small mammals, fish, eggs, insects, fruits and seeds.

Genus **Bassariscus** Coues. (Little Fox.)

Digitigrade; size rather small; body rather slender; ears large, tail about as long as head and body, annulated, skull in many respects like that of *Vulpes* in shape of teeth, shortness of palate, flatness above (not strongly arched as in *Procyon*), audital bullæ and surroundings part more like *Procyon*.

Dental formula, I. 3—3; C. 1—1; P. 4—4; M. 2—2; ×2=40.

Bassariscus astutus raptor Baird. (Sagacious; a robber)
CALIFORNIA RING-TAILED CAT.

Wood brown darkened above by black tips to the long hairs; below buffy white; basal half of the fur slate gray; grayish white area around and behind the eye and another below the ears brownish white darker on the external base; tail black above with about seven narrow white rings, these white rings widening below and taking the form of large connected triangles; soles of hind feet thickly haired half way from the heels towards the claws.

Length of male about 760 mm. (30 inches); tail vertebræ 375 (14.75); hind foot 67 (2.65); ear from crown 38 (1.50). Weight two and one half pounds. Female smaller.

California Ring-tailed Cats are not very common anywhere. In southern California they are rare. I have heard of only two

CALIFORNIA RING-TAILED CAT

instances of their being taken in the San Bernardino Mountains, and none further south. They frequent forests in the mountains. My only personal acquaintance with this species consists in trapping a pair on Eel River, Mendocino County. One of these we kept alive a few hours to observe its actions and make a drawing of it. This one permitted stroking and considerable handling, though it once nipped Mr. Fenns thumb severely when he handled it too freely.

The tracks were cat-like, not full footed like those of a raccoon. One that I got a brief look at before I put out the traps acted much like a fox. I heard a hoarse fox-like bark one night that I attributed to one of these animals. This pair inhabited a mass of boulders on a hillside thickly overgrown with brush, making a fine shelter. In dying the male emitted a weasel-like odor, not strong yet disagreeable. They are said to make nests in hollow trees. They are sometimes tamed by miners and become valued pets, as they keep the cabins free of mice. My female, caught May 16th. contained three small fœtuses. They are said to rear four kittens more often. The miners often call this animal the Civet Cat, but that name properly belongs to a different animal.

Genus **Procyon** Storr.. (Before; dog.)

Plantigrade; size rather large; body stout; ears rather small; tail about half as long as head and body, annulated; skull strongly arched; sectorial teeth modified, scarcely more than tuberculate; palate extending back about half way from last molar to audital bullæ.

Dental formula, I, 3—3; C, 1—1; P, 4—4; M, 2—2, ×2=40.

Procyon psora Gray. (Itch.)
CALIFORNIA RACCOON.

Above yellowish gray more or less heavily darkened by long black tips to the coarse hairs; underfur seal brown or hair brown;

a broad black band across the face, the eyes being within its upper border, bordered behind with grayish white which shades into the blackish of the crown and yellowish gray of the sides of the neck; region of the mouth and sides of the nose dull white with a narrow prolongation from the corners of the mouth cutting off the black facial band from a blackish patch beneath the neck; convex base of the ears and a large ill-defined spot behind them black; remainder of the ears dull white; underfur of lower parts drab gray grizzled with an intermixture of long white hairs; tail brownish buff with five to seven black or dusky rings narrower than the pale interspaces, and all but two or three of the last black rings interrupted below; tip of tail black; fore feet pale drab; toes and inner edge of hind feet pale smoke gray; remainder of hind feet sepia or dusky.

Length about 840 mm. (33 inches); tail vertebræ 295 (11.60); hind feet 125 (4.90); ear from crown 57 (2.25).

Type locality, Sacramento, California.

Raccoons are found in the timbered regions of the lower mountains and valleys of California and around some of the bays along the seacoast where no timber is near. They prefer to hunt along streams and are common in the timbered bottoms of many rivers and creeks.

The food is quite varied, including mice, small birds, eggs, insects, frogs, fish, molluscs, green corn, fruit, etc. They are good swimmers but do not dive. They are fond of fish, but can' get only such as they can snatch at the waters edge, or find dead along the shores. If they find a henroost which they can enter, or where they can reach poultry through the laths, they are likely to do considerable harm. They frequently visit vineyards, eating the ripe grapes. The harm they do is but partly offset by the considerable amount of mice that they destroy.

They travel about a considerable distance from home, sometimes having a beat several miles in length, which they take several nights to cover. I trapped one pair that got around to my vineyard about once a week, and in the interim visited one locality a mile in one direction, and another three miles in an opposite direction, a part of the route being along two small streams, and a part of the remainder along a road in a forest of scattered oaks growing among thick brush. Raccoons frequently hunt in pairs or families in the autumn, but more often alone the remainder of the year.

The ordinary gait of a Raccoon is a slow trot, and they cannot run fast. They are clever hunters, and do most of their hunting on the ground, but they pass the day in hollow trees or in crevices among rocks. They are expert climbers. They are not difficult to trap; a bait of fresh meat usually proves too much for them. The young are probably born in April and May; they are said to be three to six in number. 'Coons do not hibernate in California, the region which they inhabit not being cold enough to make hibernation necessary. I do not think they occur above 5,000 feet altitude.

Procyon psora pacifica MERRIAM.
PACIFIC RACCOON.

Similar to *psora*, but darker, the ground color being darker and the black tipped hairs very thick; black rings of tail not broken on the under side; last premolar, first molar and audital bullæ larger.

Type locality, Cascade Mountains, Washington.

Dr. Merriam says that Pacific Raccoons are common around the base of Mount Shasta. This is probably about the southern limit of this form.

Procyon pallidus MERRIAM. (Pallid.)
DESERT RACCOON.

Very pale; pattern of colors as in *psora;* buff tints of *psora* replaced by grayish white; above pale gray darkened by short black tips to the coarse hairs; below grayish white, the drab un-

derfur being nearly obscured; tail long and slender, with narrow
blackish rings; hind feet pale gray.

Length about 850 mm. (33.50 inches); tail vertebræ 310
(12.20); hind foot 130 (5.10); ear from crown 60 (2.35).

Type locality, New River, Colorado Desert, California.

Desert Raccoons are common in the bottom lands of the
lower Colorado River, frequenting the borders of the sloughs
and ponds along the overflow channels so common for miles from
the river below Yuma.   I have trapped several at the mesquit
bordered lagunas in "New River" channel in the heart of the
Colorado Desert, 50 miles from the main channel of the Colorado
River.   They follow these overflow channels, living principally
on the fish left by the overflows, helped out with birds, small
mammals, frogs and a few insects.   After the ponds dry up they
probably work back to the main channel.

Family **Ursidæ** (Bears.)

Plantigrade; size large or very large; body stout; five toes on all the feet; tail rudimentary; rostrum short; molars tuberculate; teeth 42.

This family consists of four or five genera and some fifteen or more species, widely distributed through the northern hemisphere, with one South American species They are terrestrial and most species are diurnal or nocturnal, according to circumstances. The food of most species is principally of a vegetable nature.

Genus **Ursus** LINN. (Bear.)

First three premolars in each jaw small, single rooted, often deciduous; sectorial teeth greatly modified, practically tuberculate; audital bullæ small.

Ursus horribilis ORD. (Horrible.)
GRIZZLY BEAR.

Very large; claws of fore feet very long, twice the length of the claws of the hind feet, nearly straight; skull and teeth large and massive; hair coarse; color variable, usually the tips of the hairs are yellowish or whitish in contrast with the dark basal part; general color yellowish brown, grayish or brownish yellow, with an indistinct dorsal stripe and often a dim stripe on the side; feet and legs often blackish; hairs of neck long, forming a short mane; hind legs longer than fore legs.

The length of old males is about seven feet, sometimes more, as Lewis and Clark record one measuring nine feet from nose to end of tail. The ears are about three inches long, and the tail only about two inches. The claws of the fore feet are very long, five or six inches in adult males Females are smaller than males.

Type locality, Montana.

Grizzly Bears were formerly common in California, but are

now rare, with the probability of their becoming extinct in the near future. As I have never seen a live Grizzly at large I can say nothing of their habits from personal observation. When California was first settled Grizzly Bears appear to have frequented the edges of valleys and open places in forests; but the few that are left now hide in brushy canons of rough, inaccessible parts of low mountains. They do not climb trees.

The food of Grizzly Bears is principally of vegetable nature, including roots, wild fruits, seeds, nuts, grubs and the larger insects. To this they add more or less flesh. When Grizzlies were more common they killed some domestic animals, such as colts, cattle, hogs and sheep. They are abroad as much in the daytime as in the night, appearing to hunt for food whenever they are hungry, regardless of the time of day.

Grizzly Bears are very tenacious of life, and are said to be able to run a long distance after being shot through the heart. The cubs are said to be usually two at a birth, sometimes three. They are very small when first born. Grizzlies breed readily in confinement.

It is pretty well settled now that the so-called Cinnamon Bear is a color phase of the Grizzly.

Ursus americanus PALLAS   (Black Bear.)
BLACK BEAR.

Smaller; claws of fore feet curved, not much longer than those of the hind feet; color brownish or blackish, the tips of the hairs not conspicuously lighter; pelage comparatively soft and fine.

No reliable measurements of this species are at hand.

Black Bears are still found in the northern part of California, but are rare or extinct in the southern part of the State. Their food is similar to that of the Grizzly, with a less proportion of flesh. The young are commonly two, occasionally four in number. These are born in the middle of the winter, in a very undeveloped state, and do not run about until they are about ten weeks old.

Black Bears are good climbers and swim well. They are not nearly as ferocious as the Grizzly Bears, but under all ordinary circumstances will run away from man. As their scent, hearing and sight are acute, it is difficult to get near them except by accident.

Family **Mustelidæ.** (Weasels, etc.)

Plantigrade or digitigrade; anal scent glands usually present and often highly developed: five toes on all the feet; tail long; rostrum short; sectorial teeth usually but little modified; molars usually not tuberculate; teeth 32 to 38.

The *Mustelidæ* is a large and important family, containing many species of commercial importance, and other species notable for their disagreeable odor, or for other characteristics. The family is found in nearly all parts of the world except Australia. There are fifteen or more genera, divided in three subfamilies, and nearly a hundred species. They are almost exclusively carnivorous, feeding on birds, mammals or fish caught by themselves. Some species are digitigrade, but more are plantigrade. Most species are terrestrial, some are aquatic, and a few are partly arboreal. Most species are nocturnal. Some species have very marked seasonal changes of pelage, while many wear the same colors all the year.

Subfamily **Lutrinæ.** (Otters.)

Feet webbed; body long; skull very short and wide; teeth blunt, the molars tuberculate.

Genus **Latax** GLOGER. (A sea otter.)

Fore feet small; hind feet large, fully webbed, flipper-like but haired on both surfaces; teeth comparatively smooth, massive.

Dental formula, I, 3—2, C, 1—1; P, 3—3; M, 1—2, ✕2=32.

Latax lutris nereis MERRIAM. (A daughter of Nereus, a Grecian sea god.)

SOUTHERN SEA OTTER.

Dark liver brown, with a frosted appearance, the "frosting" being caused by a scanty intermixture of long coarse hairs in the fine dense fur; head brownish white; neck grayish brown; ears very small and situated low on the side of the head; skin very loose on the body. In summer the long pale hairs are more numerous, producing a grizzled appearance.

Length about 1200 mm. (48 inches); tail vertebræ 280 (11); hind foot 150 (6) by 100 (4) in breadth.

Type locality, San Miguel Island, Santa Barbara group, California.

Sea Otters were formerly more or less common along the whole Pacific Coast from Lower California to Alaska and around to Japan. They are now rare everywhere. A very few are still living about the islands off the coasts of Lower and southern California. The fur of the Sea Otter is the most valuable of any single skin known, the price of the finest skins running up into the hundreds of dollars.

Sea Otters frequent kelp beds among rocky islets, where they feed on mussels, clams, sea urchins and other mollusks, fish and kelp. They are excessively shy, and their senses are very acute; hence they are very difficult to obtain. The single young are brought forth at any season, the intervals apparently being more than a year. The young are said to suckle more than a year.

Genus **Lutra** BRISSON.   (Otter.)

Feet short, broad, full webbed, the hind feet of normal shape; last upper premolar distinctly sectorial; tail long, tapering, not flattened.

Dental formula, I, 3—3; C, 1—1; P, 3—3; M, 1—2, $\times 2 = 34$.

Lutra canadensis pacifica RHOADS. (Of the Pacific Slope)
PACIFIC OTTER.

Dark liver brown, paler on the under side of the head, throat and breast; size averaging larger than typical *canadensis.*

Length about 1300 mm. (62 inches); tail vertebræ 460 (18); hind foot 140 (4.50).

Type locality, Kittittas County, Washington.

Pacific Otters range from Central California northward through Alaska. As the principal food of Otters is fish caught in fresh waters they do not occur in southern California. They are known to occur about suitable streams in the central and northern part of the State. At the forks of Eel River I saw where an Otter had been playing on the sand at the river shore. It had not gone more than a dozen feet from the water. Its tracks showed that it had pushed itself along on its belly in the sand.

The food of the Otter is fish caught by pursuit in the water, but usually eaten on the bank. A pastime of Otters is sliding down banks on their bellies, banks that end in the water being usually chosen. The young are born in March and April, and are one to three in number. Otters are nocturnal, very shy, and therefore seldom seen. Their fur while valuable is not nearly as high priced as that of the Sea Otter.

Lutra canadensis sonora RHOADS.
SONORA OTTER.

Similar to *pacifica,* but paler and apparently larger, postorbital processes slender; distance from point of angular process of lower jaw to summit of coronal process greater proportionally than in *pacifica.*

Type locality, Montezuma Well, Yavapai County, Arizona.

Sonoran Otters are occasionally caught in the Colorado River. While not common, they are not very rare.

Subfamily **Melinæ.** (Badgers and Skunks.)

Toes but partly or not webbed; body short and stout; skull comparatively long; sectorial teeth moderately developed.

Genus **Taxidæ** WATERHOUSE. (Badger—form.)

Body stout, very flat; tail short, flat; fore claws but little curved, very large; anal scent glands small; last molars tuberculate; audital bullæ large; palate extending half way from last molar to audital bulla; occipital crests very large; brain case triangular, very wide posteriorly.

Taxidea taxus neglecta MEARNS. (Overlooked.)
WESTERN BADGER.

Above grizzled gray, the basal half of the hairs pale yellowish brown, the tip buffy white or grayish white, the subapical fourth dusky; a narrow white stripe over the head, usually to the shoulders, often extending to the rump; nose, upper part of

Western Badger.

the head each side of the white stripe, a patch in front of the ears and feet black; under side of the head, with a point extending up between the eye and the ear white; under side of body buff, with an irregular white stripe in the middle; tail yellowish brown grizzled with black and white above, paler beneath; hairs of sides much longer than those of upper and under surfaces. *Immature;* less grizzled with gray.

Length about 735 mm. (29 inches); tail vertebræ 135 (5.33); hind foot 100 (4); ear from crown 30 (1.20).

Type locality, old Fort Crook, Shasta County, California.

Badgers are not very common in California, but are found in open country more or less throughout the State. Their food is ground squirrels, gophers, mice, eggs, insects and grubs. The only harm they do man is by digging holes in the ground, these being troublesome in cultivated ground. The young are said to be three or four in number, but I think the number must sometimes be greater, as there are eight mammæ. The young are probably born in March or April.

Badgers are principally nocturnal in habit. They are slow of foot and capture their prey principally by digging it out of burrows, for which work they are particularly adapted by their shape and great strength. They are shy and prefer hiding in burrows to fighting, but if compelled to fight they are plucky and tenacious.

There are probably two subspecies of Badgers in California, the Western (*neglecta*) in the mountains and higher valleys, and the California (*californica* Bennet) in the lower valleys. Very much more material than is now available is necessary to settle this question.

Genus **Mephitis** Cuvier. (A foul odor.)

Body rather stout; tail long, very bushy; anal scent glands highly developed; head small; skull arched; palate ending even with last molars; occipital crests large; saggital crest small; brain case not widened posteriorly; audital bullæ very small.

Dental formula, I, 3—3; C, 1—1; P, 3—3 M, 1—2, ×2=34.

Mephitis occidentalis Baird. (Western.)
CALIFORNIA SKUNK.

Pelage long and coarse, mostly black; a narrow white stripe on the crown; a broad white stripe commencing at the nape, dividing on the shoulders, running along the upper part of the sides and across the hips and ending on the sides of the tail, usually extending but a short distance on the tail—this stripe varying in form and width; hairs of tail four to seven inches in length, the basal half white and sometimes those about the middle of the tail white throughout.

Length about 685 mm. (27 inches); tail vertebræ 300 (12); hind foot 77 (3); ear from crown 22 (.85).

Type locality, Petaluma, California.

Central and northern California and southwestern Oregon, east to the Sierra Nevada and Cascade Mountains. Common in mountain and valley, timber and plain. Their food is varied, including mice, small birds, eggs, frogs, insects and grubs. Grasshoppers, beetles and their larvæ, in fact, constitute the bulk of their food when these are in season, and these Skunks are really worthy of protection for their usefulness in destroying harmful insects and mice. It is true that they do sometimes destroy poultry, but much the greater part of this damage is done by the smaller Spotted Skunks. California Skunks cannot climb, nor can they creep through very small holes, and a properly built poultry house will protect the inmates from this species.

This genus has been accused of causing hydrophobia by its bites, but there are good reasons for believing that this is a mistake, and that the bite of this species never causes hydrophobia. No cases, as far as I can learn, have been reported outside the range of *Spilogale*, and many instances are known of bites from the larger Skunks that have not resulted in an attack of the disease.

California Skunks are not as audacious as the little Spotted Skunks are; but they are very little afraid of man or beast. They are self reliant, bold and inquisitive. In spite of their powerful odor they are preyed upon by foxes, coyotes and great horned owls, as their flesh is as sweet as that of a hare or squirrel. Those persons who, not being troubled with squeamishness, have eaten it pronounce it agreeable in flavor. I never cared to try it.

Like most Skunks they have the habit, when angered or defiant, of stamping on the ground with one forefoot, or with both alternately. They are suspicious of quick movements, and act accordingly. By moving slowly one can get quite near a Skunk without provoking a discharge of scent.

The odor of a Skunk is penetrating, very pungent, and to most people very disagreeable or nauseating; when strongly inhaled it may produce unconsciousness. The fluid, much diluted and administered internally, has proved efficacious as a remedy for asthma, whooping cough and croup. Its disagreeable odor is a bar to its extended use, however. The accidental reception of a small amount of the liquid in the eye is followed by inflammation, lasting a week or so; a large amount has produced the loss of eyesight. In case of such an accident, the eye should be washed out with clean cold water as soon as possible. The odor is produced by the volatilization of a fluid secreted in a pair of glands which lie either side of the rectum, and nearly surrounding it. These glands are enveloped in a strong muscle which is capable of compressing them with sufficient force to spurt the contained fluid in several small streams a distance of twelve to fifteen feet. The ejection of the fluid is wholly within the control of the animal, and is ordinarily only resorted to in self defence.

Skunks are chiefly crepuscular and nocturnal. They occupy burrows dug by themselves when they cannot find hollow logs or suitable crevices in rocks. They are born in April, May and June, and are five to nine in number. The name Polecat belongs to an European animal of another genus.

Mephitis occidentalis major HOWELL. (Large.)
GREAT BASIN SKUNK.

Similar to *occidentalis* but larger; hind foot longer; skull larger, heavier built, broader and flattened; rostrum broader.

Type locality, Fort Klamath, Oregon.

Eastern Oregon, northeastern California, Nevada and Utah.

Mephitis occidentalis holzneri MEARNS. (For Frank X. Holzner.)
SOUTHERN CALIFORNIA SKUNK.

Similar to *occidentalis* but averaging smaller; skull narrower; teeth heavier in proportion.

Type locality, San Ysidrio Ranch, northern Lower California.

Southern California Skunks are found in northern Lower California and southern California west of the Deserts, intergrading northward with true *occidentalis*. Habits similar.

Mephitis platyrhinus HOWELL. (Broad—nose.)
BROAD-NOSED SKUNK.

Externally similar to *occidentalis;* skull short, broad, flattened in front; rostrum very broad; nasals short and broad; zygomatic arches spreading less abruptly and in an even curve nearly parallel to the axis of the skull.

Type locality, South Fork of Kern River, California.

Southern and eastern foothills of the Sierra Nevada and Owen Valley.

Mephitis estor MERRIAM. (Eater.)
ARIZONA SKUNK.

Similar to *occidentalis;* black stripe on back usually narrow; tail with white tip and more or less white at the sides, the bases of most of the tail hairs white; skull similar to that of *occidentalis;* teeth smaller; zygomatic arches heavier.

Length about 640 mm. 25.15 inches); tail vertebræ 285 (11.25); hind foot 68 (2.67); ear from crown 18 (.70).

Type locality, San Francisco Mountain, Arizona.

Arizona, New Mexico, Sonora, northeastern Lower California and eastern California along the Colorado River.

Genus **Spilogale** GRAY. (Spot—weasel.)

Body small and rather slender; tail long and bushy; anal scent glands highly developed; skull rather flat; sectorial teeth well developed; palate not extending much back of molars; saggital crest usually small; occipital crest large; auditory bullæ rather small; mastoid sinus inflated.

Dental formula, I. 3—3; C, 1—1; P, 3—3; M, 1—2. ×2=34.

Spilogale phenax MERRIAM. (Deceptive.)
WESTERN SPOTTED SKUNK.

Black with white stripes; four parallel white stripes from the head to the hips, a white stripe commencing behind the fore-

Western Spotted Skunk.

leg running back and up on the hip with a spot on each side of the backbone in the direction of its fellow stripe; a transverse stripe across the back part of the hips interrupted at the backbone; a spot each side of the rump; more or less white at the base of the tail; terminal third of the tail white, more extensive below, a white spot on the forehead; more or less white about the corners of the mouth.

Length about 400 mm. (15.75 inches); tail vertebræ 165 (6.50); hind foot 46 (1.80); ear from crown 15 (.60).

Type locality, Nicasio, Marin County, California.

Western Spotted Skunks are common in many of the valleys of central and southern California and in northern Lower California. They do not ordinarily range as high in the mountains as the larger Skunks do. The odor of the Spotted Skunk is more pungent than that of the larger species, but it is not as lasting. The only way that I know of to kill this or any other species of Skunk without its emitting its odor is by drowning. By using a box trap and carrying it to water and slowly immersing it no scent will be emitted. If a steel trap is used fasten it to the end of a long pole and the animal can be slowly dragged to the water and drowned. As long as the animal faces one there is no danger; but if it turns about stop and keep quiet until it faces about again. A Skunk will bear some pulling about if carefully handled; they do not waste their means of defense unnecessarily. Sometimes a Spotted Skunk will eat a bit of fresh meat while still in the trap, then a little strychnine will make them quiet; the meat can be reached to the animal on the end of a pole if one moves slowly and carefully.

The gait of Spotted Skunks is commonly a trot. The breeding season is about April, judging from the size of young Skunks seen in summer. They are very bold, and have so much confidence in their means of offense and defense that they seldom run from anything. Their food is much like that of the larger species—that is, mice, birds, eggs, poultry, insects and grubs. I have found parts of a snake in one's stomach. Their small size enables them to enter almost any hole that will admit a weasel or mink. These little Skunks are often very destructive of poultry, but there is another reason for destroying them; it is a well established fact that their bite does sometimes cause a form of hydrophobia. Not every bite of a Skunk will induce this disease, and one need not give up hope if bitten, yet due precautions should be taken.

Spilogale latifrons MERRIAM. (Broad—front.)
LITTLE SPOTTED SKUNK.

Colors and their pattern as in *phenax;* smaller, skull much broader in proportion to size; last molar smaller.

Type locality, Roseburg, Douglass County Oregon.

Western Oregon and northern California. Apparently not common anywhere. Habits similar to those of the Western Spotted Skunk.

Subfamily **Mustelinæ** (Weasels, etc.)

Body long; toes partly webbed; skull usually long and narrow; sectorial teeth well developed; audital bullæ usually large.

Genus **Gulo** FRISCH. (Glutton.)

Large; body stout; tail short, bushy, the hairs drooping; anal glands moderately developed; skull arched, short, wide; audital bullæ of moderate size with tubular meatus; lower sectorial tooth without inner cusp; upper sectorial tooth large; palate extending one-third of the way from last molar to audital bulla.

Dental formula, I, 3—3; C, 1—1; P, 4—4; M, 1—2, ×2=38.

Gulo luscus LINN. (One-eyed.)
WOLVERINE.

Large, blackish; an indefinite broad yellowish band on the sides, running across the hips and meeting its fellow at the base of the tail; front and sides of head grayish.

Length about 965 mm. (38 inches); tail vertebræ 200 (8); hind foot 170 (6.70).

Type locality, Hudson Bay, British America.

Wolverines are found in the colder parts of North America

and Eurasia. In California they are rare, being found only in the Sierra Nevada and the mountains of the northern coast region. I saw a mounted Wolverine in Bridgport, Mono County, that was killed in the neighborhood, and was told of others that had been killed in that region, having been driven down from the higher Sierras by winter storms. They eat anything in the way of flesh that they can capture, steal or find already dead. They are not able to run fast enough to capture many of the larger animals, and the stories told of their climbing trees and pouncing down on animals passing beneath are pure fictions, as Wolverines do not climb trees and can spring but a very short distance. Part of their food is obtained by opening the burrows of other animals, their long claws and great strength enabling them to dig rapidly. Probably a considerable part of their food in the Sierra Nevada consists of yellow-bellied marmots. They are said to be very voracious; hence their Old World name of Glutton. They are also known by the name of Carcajou.

In regions where fur trapping is carried on extensively Wolverines are a great nuisance by reason of their destroying traps and carrying away the fur-bearing animals found therein. They also have the reputation of carrying away and hiding articles for which they have no use. They are said to be very cunning and difficult to take in traps. Their scent is acute, but their sight is poor. Their fur is used for robes and trimmings. Four or five young are born in May, June or July.

Genus **Mustela** LINN. (Weasel.)

Body slender; legs short; feet rounded; claws semi-retractile; tail rather long and large; lower sectorial tooth of moderate size; palate extending about half way from last molar to audital bulla; audital bullæ rather large; occipital crest small.

Dental formula I, 3—3; C, 1—1; P, 4—4; M, 1—2, ×2=38.

July 3rd near the southwestern corner of Lassen County. Four to six young are born in April and May; perhaps as late as June. Minks are seldom found far from streams, as most of their hunting is done about water. They are fine swimmers, but poor climbers. Their food includes such birds and mammals as are ordinarily eaten by members of this family and also fish, frogs and other aquatic forms of life.

Genus **Putorius** FRISCH. (Stinking.)

Body very slender, neck long; legs very short; tail of moderate length, with rather short hairs; toes cleft; size small; anal glands moderately developed; skull flat and very narrow; upper sectorial teeth well developed, lower sectorial teeth rather small; without internal cusps; auditory bullæ large, palate extending nearly half way from last molar to auditory bulla; occipital crest moderately developed; saggital crest small.

Dental formula, I, 3—3; C, 1—1; P, 3—3; M, 1—2, ×2=34.

Putorius xanthogenys GRAY. (Yellow—under jaw.)
CALIFORNIA WEASEL.

Above cinnamon or tawny olive, tinged with yellow in summer, and with drab in winter; terminal fourth of tail black; throat, belly, inner side of legs and toes buff or ochraceous, the toes sometimes whitish; upper and lower lips, chin, sides of the head in front of and below the ears, and a large squarish spot on the forehead white, sometimes tinged with ochraceous, more often on the female; a small brown spot behind the corner of the mouth; remainder of face and top of the head varying from broccoli brown to dark sepia, darkest in winter.

Length about 420 mm. (16.50 inches); tail vertebræ 165 (6.50); hind foot 45 (1.80); ear from crown 13 (.50). Female averaging smaller.

Type locality, southern California, probably San Diego.

California Weasels are generally distributed over the valleys and lower mountains of southern California, but are common in few localities. They prey principally on mice, gophers and ground squirrels, but also eat many other species of mammals and birds. These Weasels readily enter the larger gopher burrows. It is seldom that California Weasels destroy poultry, and they should not be killed unless it is known that the individual is guilty of harmful acts, as they are highly beneficial in killing gophers and other harmful animals. Their bad reputation is partly due to the ill repute of Weasels in general, and partly to the fact that poultry killed by spotted skunks is often charged to the Weasels. A female that I caught April 18th contained six fœtuses; mammæ four pairs.

Putorius xanthogenys mundus Bangs. (Neat.)
REDWOODS WEASEL.

Similar to *xanthogenys,* but smaller and darker.

Type locality, Point Reyes, Marin County, California.

Coast region of northern California. Apparently rare; at least very few specimens have been preserved.

Putorius arizonensis Mearns. (Of Arizona.)
MOUNTAIN WEASEL.

Above raw umber or bistre darker on the head; terminal fourth of tail black; lower parts buff or ochraceous, including the fore feet, inner side of fore and hind legs, and more or less of the front part of the hind toes; chin and lips white.

Length about 380 mm. (15 inches); tail vertebræ 140 (5.50); hind foot 43 (1.70). Female smaller.

Mountain Weasels are found in the Sierra Nevada and Rocky Mountains. I do not know of their occurrence in southern or western California. I shot one at Goose Lake one forenoon as it was hunting among rocks at the base of a cliff.

Putorius muricus Bangs. (Of the mice.)
LITTLE WEASEL.

Very small; above drab brown tinged with reddish or chocolate; tail with black tip; upper lip, under parts and feet white; skull with inflated squamosals.

Length about 220 mm. (8.65 inches); tail vertebræ 60 (2.40); hind foot 31 (1.20).

Type locality, Echo, Eldorado County, California.

Order **Insectivora**. (Shrews and Moles.)

Teeth encased in enamel; upper canine, and usually lower one, present; permanent teeth rooted; lower jaw with transverse condyles received in special sockets; limbs adapted for walking, ulna and radius partly or wholly separated; metacarpal bones and phalanges of normal length; toes usually five on each foot; first and second digits not opposable; feet plantigrade or subplantigrade; placenta discoidal and deciduate.

Family **Sorecidæ**. (Shrews.)

Skull long and narrow, zygomatic arches and postorbital processes wanting; the two middle incisors of upper jaw large, curved, with a spur-like cusp at their base; lower middle incisors large and projecting forward nearly horizontally; tibia and fibula united; limbs of moderate length; feet of moderate size; the hind feet usually largest; nose elongated, tapering; eyes moderately developed; external ears present; size small or very small.

This family contains about one hundred and thirty species divided among ten genera. Individuals are most numerous in Eurasia. Shrews live in cold or temperate climates in the northern hemisphere. They are carnivorous, much of their food bebeing insects, but mice and other small animals are caught and eaten. Shrews are very ferocious animals, being able to conquer and kill mice very much larger than themselves. They are nocturnal, principally terrestrial, occasionally semi-aquatic, rarely subterranean. Seasonal changes in pelage occur in many species.

Genus **Sorex** LINN. (Shrew.)

Ears small; tail more than half as long as head and body.

Dental formula, I, 4—2; C, 1—0; P, 2—1; M, 3—3, $2 \times 2 = 32$.

The species of *Sorex* are difficult to determine. They are very small, often similar in color, with some seasonal changes of color. Some species can be distinguished with certainty only by the microscopic examination of the teeth, and to add to the difficulty these change their shape with wear.

Sorex vagrans Baird.   (Wandering.)
WANDERING SHREW.

Above dark brown, varying to almost russet; below ashy; tail dusky above, pale below; third upper unicuspid tooth smaller than fourth, fifth smaller than third.

Type locality, Shoalwater Bay, Washington.

Wandering Shrews are found from British Columbia south to the northern or probably to the central Sierra Nevada, and along the coast to San Diego.   They are found in the Transition and lower part of the Boreal Zones.   They are rare along the southern coast, but do occur in the salt marshes around tidewater bays.   Three Shrews caught on Lytle Creek, San Gabriel Mountains, San Bernardino County, seem to be Wandering Shrews, but they are very light colored, grayish sepia or hair brown.   I caught them all in July; therefore they must be in summer pelage.   They were caught in mice traps set in meadow mice runways, in grass among willows in a cool springy place, alt. 3200 feet.

Sorex amœnus Merriam.   (Attractive.)
SIERRA NEVADA SHREW.

Similar to *vagrans;* tail shorter; above dark sepia or dusky; sides paler with a gray tinge; below grayish white or buffy white; tail dark brown or dusky above, whitish below; skull and teeth similar to *vagrans.*

Length about 102 mm. (4 inches); tail vertebræ 38 (1.50); hind foot 12.30 (.50).

Type locality, Mammoth Pass, head of Owen River, California.

Higher parts of the Sierra Nevada.   They frequent wet

grassy places bordering small streams. A female caught July 22nd, contained nine fœtuses.

Sorex obscurus MERRIAM. (Dusky.)
DUSKY SHREW.

Similar to *vagrans;* larger; tail longer; ears smaller; molar teeth larger. *Summer pelage;* above dull dark sepia brown; below brownish ashy; tail dusky above, paler below. *Winter pelage;* ash gray above; whitish below.

Length about 110 mm. (4.33 inches); tail vertebræ 48 (1.90); hind foot 13 (.51).

Type locality, Salmon River Mountains, Idaho.

Dusky Shrews are found from Mount Whitney north to British Columbia and east to Colorado and Montana. They are restricted to the Boreal Zone. They are generally distributed through the higher Sierra Nevada, but have not been reported from any other part of the State. They inhabit mountain meadows and the grassy banks of streams. They often follow the runs of meadow mice and traps set in these runs sometimes catch Dusky Shrews. They sometimes eat the mice caught in traps set in the runs, and it is probable that they follow the runs partly in pursuit of mice, and partly because these runs are good hunting grounds for insects. Dusky Shrews are quite similar to Wandering Shrews in color, but they are larger, with longer skull and the molariform teeth are larger.

Sorex montereyensis MERRIAM. (Of Monterey.)
MONTEREY SHREW.

Summer pelage; above seal brown with a few long gray-tipped hairs intermixed; below light sepia; tail bicolor, sepia above, dull white below. *Winter pelage;* above slate black; below dull plumbeous brown.

Length about 120 mm. (4.75 inches; tail vertebræ 51 (2); hind foot 14.5 (.57).

Sorex pacificus Baird.

PACIFIC SHREW.

Large; hind feet large; ears large. *Summer pelage;* uniform cinnamon rufous above and below. *Winter pelage;* everywhere darker, the hairs of the upper parts dark tipped.

Length about 150 mm. (5.90 inches); tail vertebræ 63 (2.50); hind foot 17 (.67).

Type locality, mouth of Umpqua River, Oregon.

Found along the coast of Oregon and south to Point Reyes, California.

Sorex palustris navigator Baird. (Of the marsh; one who navigates.)

WATER SHREW.

Very large for a shrew; ears not conspicuous; feet with a wide fringe of stiff hairs; above slate black, some of the hairs with a short white tip producing a slightly frosted appearance; below pale brownish gray, palest on the throat and darkest on the chest; tail blackish above, dull white below except near the tip.

Length of Sierra Nevada specimens about 160 mm. (6.30 inches); tail vertebræ 76 (3); hind foot 20 (.80). Female rather smaller. Rocky Mountain specimens average smaller than those from the Sierra Nevada.

The type probably came from northern Idaho.

Water Shrews are found in the Rocky Mountains, in the interior ranges east of the Cascade Mountains from British Columbia to Utah, and in the Sierra Nevada, principally on the eastern side. They frequent the swifter mountain streams. They are strong swimmers and excellent divers, swimming under water considerable distances in the pools. They evidently obtain some of their food in the water, but I am unable to say what it is. They enter meat-baited traps. They are not very common.

Sorex bendirei MERRIAM. (For Major Charles E. Bendire.)

BENDIRE SHREW.

Large; feet with a narrow fringe of stiff hairs; ears not conspicuous; above dull sooty plumbeous; faintly paler below; tail dusky all around. Differs from *navigator* in tail being unicolor, in lower parts not being distinctly paler, and in somewhat smaller size.

Length about 150 mm. (6.15 inches); tail vertebræ 70 (2.75); hind foot 20 (.80).

Type locality, near Fort Klamath, Oregon.

Bendire Shrews occur in the Cascade Mountains from Fort Klamath to British Columbia, and along the Pacific coast from Mendocino County northward. They may occur in the mountains in the northeastern part of the State also.

Genus Notiosorex BAIRD. (Southwestern—shrew.)

28 teeth; external ear conspicuous; tail about one-third the total length.

Dental formula, I, 3—2; C, 1—0; P, 1—1; M, 3—3, ×2=28.

Notiosorex crawfordi BAIRD.

GRAY SHREW.

Above drab gray; below olive gray; tail similar.

Length about 90 mm. (3.50 inches); tail vertebræ 31 (1.22); hind foot 11 (.43); ear from crown 6.4 (.25).

Type locality, old Fort Bliss, near El Paso, Texas.

Gray Shrews seem to be rare. They are found in northeastern Mexico, in southern Lower California, in Texas and in southern California. I know of but about a dozen California examples; all were taken in dry valleys except one, which I found dead in my stable near Santa Ysabel, San Diego County, where the altitude is about 2750 feet. Two were caught near San Bernardino in fruit cans set in the ground flush with the surface.

A female caught in San Diego April 8, 1906, contained three half grown fœtuses. There were three pairs of mammæ, all located near the groins.

Family **Talpidæ.** (Moles.)

Front foot large, lateral, broad, with strong claws; hind feet normal; limbs short; no external ear; eyes minute or rudi-mentary; muzzle lengthened; body stout with no distinct neck; pelage velvety; front incisors not directed forward horizontally; zyomatic arch present.

This is a moderate sized family of about a dozen genera, generally distributed over the north temperate zone. They are carnivorous, feeding mostly on insect life obtained in burrowing through the soil. There are no obvious changes of pelage with age, sex or season.

Genus **Scapanus** POMEL. (A digging tool.)

Body spindle shaped, flattened; eyes minute, concealed in the fur but not covered by a membrane; front feet very large and broad; tail short, scantily haired, constricted at base; skull flattened; palate slightly prolonged behind last molars; first pair of upper incisors very large.

Dental formula, I, 3—3; C, 1—1; P, 4—4; M, 3—3′×2=44.

Scapanus townsendi BACHMAN. (For J. K. Townsend.)
TOWNSEND MOLE.

Very large; blackish above and below; upper unicuspid teeth separated by equal intervals; first lower incisors not much smaller than the next pair.

Length about 185 mm. (7.30 inches); tail vertebræ 40 (1.60); hind foot 25 (1).

Type locality, near Vancouver, Washington.

The Townsend Mole inhabits the region in Washington and Oregon between the Cascade Mountains and the coast range, and southwest to Crescent City, California, where specimens have been taken.

Scapanus californicus AYERS.
CALIFORNIA MOLE.

Size medium; grayish brown or light sooty brown glossed with silvery; upper unicuspidate teeth crowded and unequal in size; first pair of lower incisors very small, the next pair large.

Length about 175 mm. (5.90 inches); tail vertebræ 35 (1.40); hind foot 21 (.83).

Type locality, San Francisco, California.

California Mole.   About two-thirds life size.

Central and northern California.   They have not been reported from the San Joaquin and Sacramento Valleys and are probably rare there.   I have seen Mole runs in nearly all the mountain ranges of California, up to 5,000 feet altitude and higher.   In some parts of the mountains where the soil is of good depth and loose the runs are numerous.

Scapanus californicus anthonyi ALLEN.   (For A. W. Anthony.)
ANTHONY MOLE.

Similar to *californicus,* smaller and darker.   Brownish slate black with silvery reflections.

Length of San Diego County specimens about 155 mm. (6.10 inches); tail vertebræ 34 (1.33); hind foot 20 (.78); front foot 21 long by 16 wide (.85x63).   Female smaller.

Type locality, San Pedro Martir Mountains, Lower California.

Anthony Moles are found in the mountains of northern Lower California and in the mountains and valleys of southern California west of the Deserts. They are rare in the valleys near the coast, but are more common in the foothills and mountains, though nowhere abundant.

Moles are carnivorous. No vegetable food is eaten, the popular supposition to the contrary being erroneous. The principal food is grubs and other larvæ, insects and earthworms. They probably do not hunt for larger prey, such as mice, but a chance meeting of a mole and a mouse in a burrow would probably result in disaster to the mouse and a full meal for the mole. Mice do occasionally use mole runs for I caught a meadow mouse in a trap set in a mole run. The food is found by scent, this sense being well developed. Our moles have no visible ears, but there is a small concealed external opening. The eyes are very rudimenetary, but are not completely overgrown with skin as is the case with some other species.

The gait on the ground is very awkward, the fore feet being twisted so far outward that the Mole must walk on the thumb and edge of the foot. Our species are entirely subterranean in habit. I have never known of an instance of their coming voluntarily on the surface. Unlike various eastern and European species our Moles do not throw up mounds or "mole hills" on the surface. Their runs or burrows are often so near the surface that a narrow ridge is raised by their passage. The runs are made by the animal pressing the soil aside as it forces its way along. I captured a Mole alive and placed it in a box containing some loose soil. Its nose appeared to play an important part in burrowing. The very pliant nose was pushed in the soil and pressed to one side and the other forcing the soil aside a short distance. Into this opening the fore foot was pushed, palm outward, alongside the nose, and the foot swung outward and around, as a man swings his hand in swimming. I thought the fore feet were used alternately, but in hard soils they would probably be used simultaneously, and in hard soils the claws would probably be forced ahead of the nose to

open the way. The action of Moles in burrowing is entirely different from that of gophers (*Thomomys*), being analogous to swimming instead of digging. They burrow through loose soil very
rapidly.

Moles are sometimes troublesome in irrigated gardens
through the water following the runs. Occasionally they do a little damage by breaking the roots of plants as they force their way
along the rows of plants searching for grubs, but this damage
is usually more than offset by the benefit in destroying injurious
insects. They are very hard to trap, a special trap being necessary.
By watching where they are working they can be thrown out with
a shovel thrust in behind where the dirt is seen to move, but one
must tread lightly, for Moles are shy and their hearing is good,
notwithstanding they have no external ear.

Scapanus californicus truei MERRIAM. (For F. W. True.)

MODOC MOLE.

Similar to *californicus* but paler, clear plumbeous; rostrum
more slender; last upper premolar with a distinct inner cusp.

Length of type specimen 170 mm. (6.70) inches); tail vertebræ 34 (1.33); hind foot 21 (.83).

Type locality, Lake City, Modoc County, California.

Genus Neurotrichus GUNTHER. (New—tail—hair.)

Body spindle shaped; eyes small, concealed in fur, not covered by a membrane; front feet moderately broad; tail about half
as long as head and body, thinly haired, constricted at base; skull
flattened; palate ending even with last molars; first pair of upper
incisors moderately large.

Dental formula, I, 3—3; C, 1—1; P, 2—2; M, 3—3'×=36.

Neurotrichus gibbsi major MERRIAM.   (For George Gibbs; large.)

LARGE SHREW MOLE.

Dark sooty brown with purple and silvery reflections.

Length about 120 mm.   (4.75   inches) ;   tail vertebræ 40 (1.57) ; hind foot 17 (.67).

Type locality, Carberry Ranch, Shasta County, California.

Northern Sierra Nevada and Mount Shasta above 4,000 feet altitude and the coast region north of San Francisco.   Not common.

Order **Chiroptera**. (Bats.)

Fore limbs modified for flight by the elongation of the forearm and fingers: fore and hind limbs connected by a membraneous
expansion of the skin, this frequently including the tail: humerus
and femur extending beyond the body; bones of the   forearm
united; ulna reduced to a rudiment: hind limbs so far rotated that
the knee bends outward and backward; a cartilagineous calcar on
the inner side of the ankle of the hind foot supporting a part of
the interfemoral membrane: teeth enveloped in enamel and consisting of incisors, canines, premolars and molars.

The highly specialized order of Bats is widely distributed
over the globe excepting in the polar regions. The order consists of two suborders and six families. One suborder (*Mega-
chiroptera*) does not occur on this continent. Its members feed
principally on fruit. Some species are very large, such as the so-
called Flying Foxes, some of which are as large a as large hawk,
while other species are quite small.

The wings of Bats consist of a web-like expansion of the
skin from the upper and lower surfaces of the body, these two
layers being thin, coherent and expanded by a framework consisting of the greatly lengthened bones of the fingers and arms and
the more or less lengthened and exserted legs; the membrane being continued from the end of the inner finger to the foot of the
hind leg and usually to the tail. The flight of a bat is not as
graceful as that of a bird, but it is nearly as rapid and more completely under control in making rapid turns. They are as awkward in walking on the ground or other surfaces as they are dex-
trous on the wing.

The eyes of bats are small and of less service than most other
of their senses. The organs of smell are well developed. The
sense of touch or feeling is highly developed, especially in the
wing membranes and nasal appendages of the "leaf-nosed"
species. The hearing is very acute and is probably the most useful sense in locating their insect prey.

Bats are crepuscular and nocturnal, rarely going abroad in

daylight. Some species spend the day in narrow crevices, into which they crawl, sometimes in large numbers; other species hang from the roof of caves, often in masses; yet others hang in trees from twigs among the foliage. Very little is known about the migrations of bats, but there are very good reasons for believing that many species migrate in a method similar to that of birds. Probably few species occuring in cool climates remain there in winter.

The number of young at a birth is commonly one or two; rarely three, so far as is known. With certain species one young may be the rule, with many two is the usual number. Most have but one pair of mammæ, but others, as *Lasiurus,* have two pairs. It is probable that some species rear two sets of young annually. Many species are gregarious, but usually the two sexes do not intermingle.

Suborder **Microchiroptera.**

Insectivorous bats of medium or small size; molars with crowns acutely cuspid.

Family **Vespertilionidæ.**

Upper incisors small, with a vacant space in their middle; molars with conspicuous W-shaped cusps; turbinal bones folded; tail included nearly to tip in the interfemoral membrane; ears medium or large, usually well separated; tragus well developed; no distinct nose leaf; hairs surrounded with minute imbricated scales.

This family of Bats contains seventeen genera and one hun_dred and fifty or more species, most common in temperate cli_mates. The sexes are alike. The young differ but little from the adult. There are no seasonal changes of pelage.

Genus **Antrozous** ALLEN. (Cave—animal.)

Ears not joined at base; muzzle blunt; lower lip free.
Dental formula, I, 1—2; C, 1—1; P, 1—2; M, 3—3×2=28.

Antrozous pallidus LECONTE. (Pallid.)
PALE BAT.

Size large; ears large; tragus slender, nearly straight, a little less than half as high as the ear conch; interfemoral membrane of moderate size; wings broad; back pale drab gray, most of the hairs with faintly dusky tips; below grayish white, tinged with drab on the sides.

Length about 110 mm. (4.33 inches); tail vertebræ 40 (1.60); ear from crown 25 (1).

Type locality, El Paso, Texas.

The pale Bat is found from western Texas through the arid region of the Sierra Nevada and San Bernardino Mountains. They do not seem to be common anywhere.

Pale Bat.

Antrozous pallidus pacificus MERRIAM.
PACIFIC PALE BAT.

Averaging larger than *pallidus;* darker; above brownish white more or less heavily tipped with sepia or drab, a patch on the back of the neck and sometimes one on the rump with little or no dark tips to the hairs; below buff or brownish buff.

Type locality, old Fort Tejon, California.

Pacific Pale Bats appear to be generally distributed along the Pacific coast west of the Cascade Mountains and Sierra Nevada from the Columbia River south to Cape St. Lucas, in the valleys, foothills, and lower mountains. They do not appear to be common. The young are born about the first of Ju .

Genus **Euderma** ALLEN. (Beautiful—skin.)

Ears enormous, joined together at their bases by a low membrane across the crown; tragus joined to external lobe of ear; tip of ear rounded; face without evident glandular swellings.

Dental formula, I, 2—3; C, 1—1; P, 2—2; M, 3—3×2=34.

Euderma maculatum J. A. ALLEN. (Spotted.)
SPOTTED BAT.

First upper premolar minute; ears marked with numerous transverse lines; nose without a leaf or other excrescence; face thinly haired; color peculiar in being distinctly spotted; base of ears and upper sides of neck whitish; a spot on each shoulder and one on the rump white at tips and black at base of hairs; remainder of fur on back dark sepia; fur of under part of body black at base and white at tips.

Length about 110 mm. (4.33 inches); tail vertebræ 50 (2); ear from crown 43 (1.70).

Type locality, Castac Creek, Los Angeles County, California.

But three specimens are known of this peculiar species. The type was found hanging on a fence; the second specimen was found dead in the Biological Laboratory of the New Mexico College of Agriculture at Messilla Park, New Mexico; and Herbert Brown reports the capture of another at Yuma, Arizona.

Genus **Corynorhinus** ALLEN. (Club—nose.)

Ears very large, thin, joined together over the crown, the back half of the ear with numerous transverse lines; tragus slender, straight, notched and lobed near the bottom, about two fifths as long as the ear; an upright glandular mass each side of the face between the nostril and the eye; interfemoral membrane large.

Corynorhinus macrotis pallescens MILLER. (Very pale.)
LUMP-NOSED BAT.

Above yellowish sepia, the bases of the hairs tingled with plumbeous; below yellowish drab or pale drab; ears and membranes light brown.

Lump-nosed Bat.

Length about 98mm. (3.85 inches); tail vertebræ 48 (1.90); ear from crown 30 (1.20); expanse of wings 285 (11.25).

Type locality, Keam Canyon, Navajo County, Arizona.

Lump-nosed Bats are found in the deserts, valleys and foothills of California and eastward to Colorado and Texas. They are common. They are summer residents in this State, but probably a few winter in warm localities. I have a specimen taken at San Diego in March. Another taken April 25th, contained one fœtus. They are on the wing before the twilight is gone. They appear to inhabit caves.

Genus Myotis KAUP.. (Mouse—ear.)

Face hairy; muzzle and nostrils simple; ears not connected at base; interfemoral membrane ample.

Dental formula, I, 2—3; C, 1—1; P, 3—3; M, 3—3×2=38.

Myotis lucifugus longicrus TRUE. (Light—fugitive; long—shank.)
LONG-SHANKED BAT.

Above varying from sepia to yellowish black; below varying from pale hair brown to sepia; membranes dusky or blackish; ears rather small, broad, upper part of back edge concave; no fringe of hairs on border of interfemoral membrane; tibia proportionally long.

Length about 97 mm. (3.80 inches); tail vertebræ 42

(1.65) ; ear from crown 14 (.55) ; expanse of wings 275(10.80).

Type locality, Puget Sound.

Long-shanked Bats range over much of the western United States, but are common in few places. In some parts of their range they inhabit mountains, but most of the recorded specimens were taken in valleys.

Myotis californicus Audubon and Bachman.
LITTLE CALIFORNIA BAT.

Size small; feet small; ears small, reaching just beyond tip of nose when laid forward; back edge of ears concave; no fringe of hairs on border of interfemoral membrane; color above reddish sepia or drab, below a paler shade of the same color; fur everywhere blackish at base; membranes dull brown or dusky.

Length about 82 mm. (3.25 inches) ; tail vertebræ 40 (1.60) ; ear from crown 13 (.50) ; expanse of wings 230 (9).

California Bats range in the valleys, foothills and lower mountains of the coast region of the western United States and in Lower California. They are common in the valleys of California in the autumnal migration and are present in smaller numbers all summer. They hide in the daytime in crevices in rocks, behind loosened boards in barns and other buildings and in other dark crannies, coming out in early twilight.

Myotis californicus pallidus Stephpns.
PALLID BAT.

Averaging smaller than *californicus;* paler; above buff or brownish buff; below dull white; all pelage dusky at base.

Length about 80 mm. (3.15 inches) ; tail vertebræ 40 (1.60) ; ear from crown 11 (43) ; expanse of wings 210 (8.25).

Type locality, Vallecito, San Diego County, California.

Pallid Bats are found in summer in the Colorado and Mojave Deserts and in the arid mountains around them. A female taken April 29th, contained one small fœtus. A few Bats winter in the Colorado Desert; these appear to be intermediate between *pallidus* and *californicus*.

Myotis yumanensis ALLEN.   (Of Yuma.)
YUMA BAT.

Similar to *californicus;* lighter color; body larger; tail shorter; hind foot much larger; skull broader.

Type locality, old Fort Yuma, California.

Yuma Bats are found in the southwestern United States and northwestern Mexico. They appear to be most common in the San Joaquin Valley.

Myotis yumanensis saturatus MILLER.   (Full of color.)
MILLER BAT.

Similar to *yumanensis;* darker colored; smaller; back dark glossy yellowish brown; belly isabella color; fur nearly black at base.

Type locality, Hamilton, Washington.

Miller Bats are found in British Columbia, Washington, Oregon and northern California. Dr. Merriam reports them com mon high on Mount Shasta in August.

Myotis evotis ALLEN.   (Good—ear.)
LONG-EARED BAT.

Ear very long for this genus, narrow; size rather large; no fringe of hairs on the border of the interfemoral membrane; wings rather narrow; above wood brown or isabella brown; below pale drab; fur everywhere blackish at base; wings and ears dark brown.

Length about 90 mm. (3.55 inches); tail vertebræ 41 (1.60); ear from crown 21 (.82); expanse of wings 240 (9.50).

Type locality, Monterey, California (Miller).

Long-eared Bats are found in the western United States

and Mexico. They are not abundant. In southern California I have seen this species most frequently in the spring and fall migrations. They are abroad in twilight. They frequent both mountain and valley.

Myotis thysanodes MILLER.   (Fringe—like.)
FRINGED BAT.

Size medium; border of interfemoral membrane thickened from end of calcar to tip of tail and distinctly fringed with hairs; ears rather long, reaching three to five millimeters beyond the nostrils when laid forward; feet rather large; above dull yellowish brown; below a paler shade of the same color; fur everywhere blackish at base.

Length about 90 mm. (3.55 inches); tail vertebræ 37 (1.45); ear from crown 17 (.67).

Type locality, old Fort Tejon, California.

Fringed Bats are known only from southern California and northwestern Mexico. They appear to be common in the type locality, where Dr. Merriam and Dr. Palmer found them hanging in clusters from the rafters in the attic of an old building forming part of the abandoned quarters of the old Post, in company with Yuma Bats. Young of various ages were found with the adults July 5th, 1991.

Genus Lasionycteris PETERS.   (Hairy—bat.)

Skull flat; rostrum broad; face mostly bare and glandular; ears low, broad, widely separated; tragus short, broad, straight in front, convex behind; basal half of interfemoral membrane furred on the upper side.

Dental formula, I, 2—3; C, 1—1; P, 2—3; M, 3—3, ×2=36.

Lasionycteris noctivagans LE CONTE.   (Night—wandering.)
SILVERY-HAIRED BAT.

Above and below blackish chocalate brown tipped with silvery white.

Length about 100 mm. (3.95 inches) ; tail vertebræ 40 (1.60) ; ear from crown 15 (.60).

Type locality, eastern United States.

Silvery-haired bats are common in the eastern United States, but appear to be rare west of the Rocky Mountains. I have seen no Californian examples and know of but eight having been taken in the State. In the eastern States this species frequents the vicinity of streams and the borders of hardwood forests.

Genus **Pipistrellus** KAUP. (A bat.)

Size small; skull small and lightly built; ears longer than broad, tapering to a narrow rounded tip; tragus straight or curved forward; basal fourth of interfemoral membrane thinly haired on the upper side.

Dental formula, I, 2—3 ; C, 1—1 ; P, 2—2 ; M, 3—3×2=34.

Pipistrellus hesperus ALLEN. (Western.)
WESTERN BAT.

Smallest California species of bat; ear short, barely reaching nostril when laid forward; ears widely separated; tragus rather short, very blunt and bent forward; feet small; interfemoral membrane of moderate size, sparsely haired on the upper surface near the body, the border not fringed; face and ears bare, black; color of pelage pale; above very pale drab; below brownish white; all the fur blackish at base; wings dull black.

Western Bat.

Length about 72 mm. (2.85 inches) ; tail vertebræ 30 (1.20) ; ear from crown 10 (.40) ; expanse of wings 200 (7.90).

Type locality, old Fort Yuma, California.

Western Bats range from southern and eastern California east to Colorado and Texas. They are a desert loving species and are not common in the coast region of southwestern California. Very few remain in California in winter. The northward migration is at its height about the end of March, at which time they are very abundant about certain springs along the western border of the Colorado Desert, appearing early in the evening, sometimes soon after sunset. By the middle of April they are much less abundant about these springs. Their flight is swift and erratic and they are hard to shoot. They probably hide in crevices in rocks on hillsides during the daytime. I found two fœtuses in a female shot May 18th.

Genus **Eptesicus** RAFINESQUE. (House flier.)

Skull large and heavily built; size rather large; ears rather short and narrow; tragus rather short, narrow, pointed; wing and tail membranes naked; wings large.

Dental formula, I, 2—3; C, 1—1; P, 1—2; M, 3—3×2=32.

Eptesicus fuscus bernardinus RHOADS. (Brown; of San Bernardino.)

SAN BERNARDINO BAT.

Above wood brown or isabella brown; below paler; skull flat; rostrum very broad.

Length about 110 mm. (4.33 inches); tail vertebræ 46 (1.80); ear from crown 14 (.55); expanse of wings 330 (13).

Type locality, San Bernardino, California.

Southern California, principally in the mountains. Rather common in summer in the pine region.

Eptesicus fuscus melanopterus REHN. (Black—wing.)

SIERRA BAT.

Similar to *bernardinus* but darker; above dark cinnamon; below reddish wood brown; face and membranes black.

Type locality, Mt. Tallac, Sierra Nevada, California.

The range of the Sierra Bats has not been worked out, but it is probably all the forested region of central and northern California and perhaps all the west coast region north of California also.

Genus **Lasiurus** Gray. (Hairy—tail.)

Skull very short, broad, high; but one pair of upper incisors, divided by a wide space; first upper premolar minute, crowded out on the tongue side of the canine; upper side of the interfemoral membrane furred to the edge; ears broad, low, more or less furred; tragus rather short, curved; mammæ four.

Dental formula, I, 1—3; C, 1—1; P, 2—2; M, 3—3×2=32.

Lasiurus borealis teliotis Allen. (Northern; perfect— ear.)
WESTERN RED BAT.

Ears low, broad, the side toward the crown thickly furred, the outer side with a few scattered hairs; tragus short, pointed, wide, strongly curved; wings furred next the body on both sides and on the under side a thin strip of fur one fourth the width of the wing extends to the wrist; under side of interfemoral membrane bare except near the base, upper side middle half of hairs buffy or pale yellowish, tips a reddish shade varying from tawny or cinnamon to ochraceous buff, sometimes thinly frosted with white; below pale ochraceous or yellowish; fur of upper side of interfemoral membrane mostly reddish throughout.

Western Red Bat.

Length about 110 mm. (4.33 inches); tail vertebræ 50 (1.95); ear from crown 6 (.23); expanse of wings 315 (12.40).

Type locality, California.

Western Red Bats are found in the valleys and foothills of central and southern California and Lower California. All that I have seen were found in spring and summer hanging among the foliage of fruit trees in orchards. They appear to be rare.

Lasiurus cinereus Beauvois.   (Ashy.)

HOARY BAT.

Large; ears mostly furred on both sides; a spot of fur on the upper side of wing near elbow and one or two at wrist; a strip of fur on the under side from elbow to wrist; upper side of inter-femoral membrance thickly furred; under side bare except near body; upper pelage blackish at base, the middle of the hairs pale yellowish brown becoming umber brown on the interfemoral membrane, tips distinctly hoary white with a narrow chocolate sub-terminal zone; head mostly ochraceous; breast and much of the belly similar to the back; remainder of lower parts, including throat, grayish buff.

Length about 135 mm. (5.30 inches); tail vertebræ 57 (2.25); ear from crown 13 (.50); expanse of wings 400 (16).

Type locality, Philadelphia, Pennsylvania.

Hoary Bats are found in most parts of North America. In summer they mostly frequent mountains or cool hilly regions. Several have been found hanging in the thick foliage of orange trees in southern California in winter. I found them in May in the redwoods of Mendocino County. Their flight is swift, with frequent abrupt turns. They do not appear until the light becomes very dim.

Family **Molossidæ.**

Upper incisors large, separated by a vacant space in the middle in some species, not separate in others; molars with distinct W-shaped cusps; wings narrow; terminal third or half of tail vertebræ free and projecting beyond the narrow interfemoral membrane; ears medium or large, usually separated at base; tragus more or less developed; no nose leaf; scales on hairs arranged in belts.

This family is principally tropical or subtropical in distribution. It contains half a dozen genera and about fifty species.

Genus **Nyctinomops** MILLER. (Night—habitation—like.)

Size medium; a space between upper incisors; first upper premolar very small; membranes not furred; lips large and thick.

Dental formula, I, 1—2 or 1—3; C, 1—1; P, 2—2; M. 3—3 ×2=30 or 32.

Nyctinomops mohavensis MERRIAM. (Of Mohave.)
MOHAVE BAT.

First upper premolar minute; third lower incisor minute, sometimes lacking in adult or aged individuals; front border of

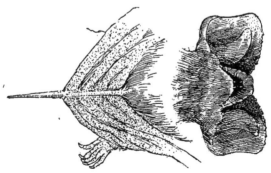

. Mohave Bat.

ear with about six wart-like small projections; numerous black spines scattered over the face and chin; lips crimped, forming perpendicular wrinkles; ears apparently connected at bases, but usually not really united; tragus small; free part of tail about equal to part included in the membrane; wings narrow; color above sooty mouse gray; below smoke gray; membranes dark brown.

Length about 98 mm. (3.85 inches); tail vertebræ 37 (1.45); ear from crown 13 (.50); expanse of wings 310 (12.20).

Type locality, Fort Mohave, Arizona.

Mohave Bats have been taken in various parts of Arizona and California. It is probable that they occur over most of the southwestern United States and northwestern Mexico. Stowell found them in large numbers in the courthouse at Santa Clara, California in February. I have taken them on the borders of the Colorado Desert in March and April, and at San Diego in November. I am under the impression that this species migrates, but this is not yet proven to be a fact. Nearly all that I have seen or heard of were taken in valleys, but probably a few get into the lower mountains in summer. They seem to feed mostly on species of insects that fly over water and damp places. They spend the day in crevices of rocks, behind shutters and in cracks of buildings, sometimes in masses. They begin to fly rather early. The flight is erratic but not swift.

Nyctinomops femorosaccus MERRIAM. (Thigh—sack.)
POCKETED BAT.

Similar to *mohavensis;* larger; tail more than half exerted; a fold of membrane extends from the inner third of the femur to the middle of the tibia, forming a pocket at the thigh; ears connected at the base; color dull brown.

Length (type) 103 mm. (4.05 inches); tail vertebræ 41 (1.60); free part of tail 23 (.90); ear from crown 14 (.55).

Type locality, Agua Caliente (now called Palm Springs), in the northwestern end of the Colorado Desert, California.

I shot the type specimen March 27th, 1885, but have not recognized more of the species since, and have seen no records of further captures.

Nyctinomops depressus Ward. (Depressed.)
NEVADA BAT.

Size large; two pairs of lower incisors; ears united at their bases; above dull brown; below similar but lighter; males with a small sac in the skin of the throat.

Length about 140 mm. (5.50 inches); tail vertebræ 41 (1.60); expanse of wings 410 (16.15).

Type locality, Tacubaya, Federal District, Mexico.

The Nevada Bat is found in Mexico and the southwestern United States. It must be rare in the United States as the only records that I can find are one each for California, Nevada, Arizona and Colorado. It should be readily distinguished by its large size.

Genus Promops Gervais (Before—Mops.)

Size large; no space in the middle between the upper incisors; first upper premolar very small; lips not wrinkled; ears united at base; membranes not furred.

Promops californicus Merriam.
CALIFORNIA MASTIFF BAT.

Very large; first upper premolar minute and wedged in the angle between the canine and the second premolar on the outer side of the tooth row; ears broad, projecting a little beyond the nostrils when laid forward; tragus quadrate, higher than broad; a glandular swelling in front of each eye; color sooty brown, paler below, the bases of the hairs everywhere pale drab gray.

Length about 162 mm. (6.38 inches); tail vertebræ 60 (2.35); free part of tail 13 (.50).

Type locality, Alhambra, Los Angeles County, California.

California Mastiff Bats are rare. They are known only from southern California. They have been found over a door, behind a signboard, hanging from a window ledge and in a tunnel. All dates known to me are in winter.

Family **Phyllastomatidæ.** (Leaf-nosed Bats.)

Upper incisors not separated by a space in the middle; but four lower incisors; cutaneous processes present about the nose or mouth; ears medium or large sized; tragus developed.

This family is confined to America and is numerous in species in the tropics. Some species eat fruits as well as insects.

Genus **Otopterus** Lyddeker. (Ear—wing.)

Nose leaf simple, erect; ears large, united at base; point of tail extending beyond the interfemoral membrane; skull long and slender.

Dental formula, I, 2—2; C, 1—1; P, 2—3; M, 3—3×2=34.

Otopterus californicus Baird.

CALIFORNIA LEAF-NOSED BAT.

Nose with an upright "leaf" of cartilege and skin; ears very large, connected at their bases; tragus slender, pointed, one-third

California Leaf-nosed Bat.

the height of the ear; wings broad, not furred; interfemoral membrane small, concave in outline; basal. half of pelage white, outer half broccoli brown, darkest above, paler beneath, slightly tipped with white; membranes light brown.

Length about 95 mm. (3.75 inches); tail vetebræ 41 (1.60); ear from crown 28 (1.10); expanse of wings 330 (13).

Type locality, old Fort Yuma, California.

California Leaf-nosed Bats are found in Southern California, Arizona, western Mexico and Lower California." In California they frequent valleys and foothills. They are probably migratory.

I know of no instance of their occurance in California in winter, and I have failed to find them at all in January in a place where I can nearly always find them in spring and summer. They probably spend the day in caves, crevices in rocks and similar dark places. I have not seen them on the wing until all the twilight has faded away. The young are born in June. More than half of the females bear two young, the remainder but one. ⸗

Order **Primates**.

Inner digit of hand, and in some families the inner digit of foot, opposable to the other digits; femur and humerus fully exserted; clavicles present; orbits encircled by bone and directed forward.

Family **Hominidæ**. (Man.)

Body erect; inner digit of foot not opposable to the other digits; five digits on each limb; cranium large; cerebral hemispheres of brain very large; canine teeth but moderately developed; tooth row without gap; no tail vertebræ; hair developed only on special areas; ears rounded, with a soft dependent lobule.

If the same rules of classification be applied to Man that are applied in the lower orders he must be included in the *Primates* with monkeys, apes, etc., but placed in a family separated from them by characters that, taken together, show a higher organization. "The essential attributes which distinguish Man and give him a perfectly isolated position among living creatures are not to be found in his bodily structure." They are mental, not physical, and zoological classification is based only on physical characters.

Using terms similar to those used in preceding families we may say that the *Hominidæ* are distributed over all parts of the land surfaces of the earth; they are plantigrade; terrestrial; diurnal and crepuscular; omnivorous; more or less gregarious; the adult males differ somewhat from the females; the immature are similar to the female; mammæ two, pectoral; there are usually but one young at a birth, occasionally two, rarely more; the young develope slowly.

The *Hominidæ* contains but one genus, which is considered by most zoologists to be composed of a single species, divided in several races (technically subspecies) which blend so thoroughly at one point or another as not to be separable into distinct species. These races vary greatly in physical and mental qualities.

Genus **Homo** LINN. (Man.)

Facial angle high; arms shorter than legs; nail flattened, present on all the digits.

Dental formula, I, 2—2; C, 1—1; P, 2—2; M, 3—3×2=32.

Homo sapiens americanus LINN.
AMERICAN INDIAN.

Hair coarse, round in transverse section, straight, black, long and abundant on the scalp but sparse elsewhere; skin dark, often with a bronzy tinge; forehead retreating; nose prominent, usually with a high bridge; eyes horizontal.

When the first Europeans came to California the Indians were numerous and distributed over all the State except the higher parts of the mountains and the waterless deserts. Excepting the desert tribes the California Indians were a quiet peaceable people. They had few vices but were very superstitious and somewhat revengeful. They were humble, contented and industrious considering the ease with which their few natural wants could be supplied. Unlike the Indians of eastern North America they did not torture prisoners nor scalp slain enemies. The chiefs or head men had very little real authority, the conduct of individuals being guided mostly by old customs and superstitions. The tribes were small and weak. In his report on the "Tribes of California" (1877) Powers names over one hundred and fifty tribes, a few of these being small sub-tribes, the remainder being independent and speaking distinct dialects. He did not include the Mission Indians, nor the Indians of the Colorado valley.

It has been found that one of the best clews to the relationship between human races is their languages. This seems particularly true of American Indians. The polysynthetic feature of speech runs through all the various American languages and dialects. Their construction is radically unlike that of Eurasian languages and seems to point to a separation from the people now inhabiting Europe and Asia soon after the acquirement of language by the human races.

Few regions of similar extent to California can show a greater number of contemporaneous native dialects. This seems to have been the results of a very early immigration and settlement of small tribes or fragments of tribes combined with a strong home-loving trait, which may have been a late development. When found by the whites the Indians did not seem to care to travel and mingle with their neighbors and each tribe and often each community had a dialect of its own. According to the latest map of the Bureau of Ethnology these dialects were grouped in twenty one linguistic stocks, the total number of North American stocks being about sixty-five. The names and distribution of the California linguistic stocks are given on the accompanying map.

The question of the derivation of the race of American Indians has interested many persons. Where the original cradle of the human race was, probably will never be positively decided. It is usually supposed to have been in Asia, yet it may possibly have been in America. It is probable that the dispersion of races occured in pre-glacial times from a well populated circumpolar region. Glaciation slowly forced the inhabitants from the polar regions toward the tropics, and the cold and ice separated the inhabitants of America from those of Europe and America. The difference in construction of the languages of the two continents indicates that this separation occured in a very early stage of language formation. With the melting of the ice and the retreat of the glaciers, which is still progressing, the tribes of Indians nearest the vacated region were able to move slowly northward, allowing other tribes to expand or follow if they chose. The high Sierra Nevada range, being heavily capped with snow and ice, was an impassable barrier between the interior of the continent and the comparatively warm coast region of California. Probably fragments of migrating tribes were forced through the passes before these became impassable, and could get no further. When the glaciers retreated the California Indians did not follow, as there was no pressure from beyond and no inducement to leave.

Until the rush of gold seekers no great change occurred in

California Linguistic Stocks

1. Athabascan.
2. Yurok.
3. Karok.
4. Shasta.
5. Lithuami.
6. Wishosk.
7. Chimariko.
8. Wintun.
9. Yana.
10. Maidu.
11. Yuki.
12. Pomo.
13. Washo.
14. Moquelumnan
15. Costanoan.
16. Esselen.
17. Yokuts.
18. Salinan.
19. Chumash.
20. Shoshonean.
21. Yuman.

the number of Indians in California, but a rapid diminution then began; partly through the unjustifiable persecution by the stronger, better armed, aggressive gold-seekers, many of whom cared nothing for the moral rights of the Indians; partly through the introduction of intoxicating liquors; but more through the effects of epidemic and other diseases which came with the whites. Now some of the smaller tribes are practically extinct, but under more wholesome conditions the younger generation seems to be nearly holding its own or slowly increasing in a few places. The California Indians seemed to lack the power of organization and the faculty of invention, hence they made little progress toward civilization until the whites came and took the lead. Their recent progress shows that they are capable of considerable education.

The name Amerind has been proposed for the native races of America. It is composed of the first syllables of America and Indian.

Life Areas of California

Most people who have ascended mountains, on business or for pleasure, have noticed that there was a gradual change in the trees and other vegetation as height was gained, and some see that there is a system in this change. At a certain height in one mountain occurs a combination of trees, shrubs, plants, birds, insects and mammals, which combination is repeated in a general way on other mountains at a similar altitude, modified by local causes, such as soil, angle or direction of slope, nearness or remoteness of large bodies of water, height above base level and other conditions. Going higher, a change in the birds, trees, etc., occurs through the gradual disappearance of some species and the substitution of others until a new combination is formed. A similar combination is repeated in other mountains of the region in about the same order. Local causes modify these repetitions more or less, but the general similarity is sufficient to force the close observer to the conclusion that they are controlled by general natural laws. Within a few years much study has been given to the elucidation of these natural laws, and I will attempt to summarize some of the results of these investigations in California.

The causes controlling the geographical distribution of life are many, the most important being temperature, moisture, soil and light. We are accustomed to sum up three of these leading causes in the word climate.

The most important single cause of the varied distribution of life is heat; its quantity and daily and yearly range over a given area. Other conditions being equal, the warmer the climate of a locality is, the more luxuriant and varied it forms of life will be. A great yearly or daily range of temperature unfavorably affects the life of an area by weeding out the forms most sensitive to such changes, on the principle of the "survival of the fittest."

The heat of a locality is affected by its latitude, altitude, direction of the prevailing winds, height above base level and slope exposure. Increase of latitude and altitude produce similar climatic effects, the higher area having a similar climate to that of the lower area situated a certain distance further from the equator. In other words, a traveler passing from the tropics toward the poles at sea level finds the climate steadily becoming colder; in climbing a mountain the same change is observed.

If the area of high altitude is great it is warmer than a small similar area at the same height and latitude, for the reason that the greater area conserves the greater amount of heat as daily received from the sun. It sometimes happens that the base level on one side of a mountain range is higher than that on the other side; in this case the higher level tends to raise the temperature and therefore the life zones on that side. A good illustration is the Himalaya Mountain range. The plain on the south side is several thousand feet higher than the plateau on the north side; in consequence of this difference of base level on the two sides the timber line and snow line are about three thousand feet higher on the north than on the south side. This is in direct opposition to the effect of latitude which would tend to lower the snow line on the north side. The Sierra Nevada Mountains are another illustration. The plateau on the eastern side is from three to four thousand feet higher than the San Joaquin and Sacramento Valleys on the west side, and in consequence all the life zones are higher on the east side than on the west.

Slope exposure is another disturbing cause. A slope directly facing the sun is warmer than one facing away from it. This is very noticeable in many canyons running east and west in semiarid parts of California, in which case the timber will be found growing considerably lower down on the side receiving the least amount of direct sunshine.

Prevailing winds coming directly from large bodies of water tend to cool the region contiguous and therefore lower the life zones.

The next most important agent in the distribution of life is moisture. The greater or lesser amount of moisture present in air and soil strongly affects the vegetable growth of a locality; as

animal life of a locality is practically dependent on the vegetation
it is in that way affected by the proportion of moisture present.
The amount of moisture of a region is regulated by its distance
from large bodies of water, the direction of the prevailing air currents, and the height of intervening obstacles, such as mountain
ranges. Most of the moisture present in the air originates in the
evaporation of seas and other large bodies of water. The moisture
laden air moving inland when cooled is unable to hold up all its
moisture, which falls as rain. A high range of mountains will
greatly cool the air currents passing over it and the heavy rainfall
or snowfall resulting may abstract so much of the moisture from
the air, that little is left for the region beyond the mountains,
which thus becomes arid. The region of the Colorado and Mojave Deserts and the greater part of Nevada is an illustration of
the drying influence which the Sierra Nevada Mountains exert
on the air currents passing over them.

The quality of the soil is another factor in the quantity and
character of the plant and animal life of a region. The carnivorous species of animals of a region subsist on the herbivorous species; these subsist on the leaves, stems, seeds or root of plants
which draw their nourishment from the soil; therefore a richer
or poorer soil has a considerable direct influence on such apparently remotely connected beings as the foxes or hawks that live in a
region.

Dr. C. Hart Merriman has formulated certain laws of the distribution of life which appear to be based on sound reasoning from
a sufficient mass of observed facts to assure their correctness.

"The northward distribution of animals and plants is determined by the total amount of heat—the sum of effective tem-
peratures.

The southward distribution of Boreal, Transition zone, and
Upper Austral species is determined by the mean temperature of
the hottest part of the year."

If the North Temperate Realm was composed of sea and
level land only, its life zones would nearly follow parallels of

latitude around the northern hemisphere, deflected here and there by the effects of warm or cold ocean currents on the shores they wash. The presence of mountain ranges breaks up such uniformity of climate and renders the definition of life zones very difficult, nowhere more so than in California, where, in many mountains, island-like areas are detached from the main bodies of their zones or long points project, or narrow bands curve to follow the sinuosities of the mountain sides. The peculiar topography of this state produces a variety of life zones which is probably equaled by no other similar area elsewhere. Bordered as California is by the sea; traversed its whole length by a mountain range, in places carrying perpetual snow; possessing considerable areas lying below sea level; having a range of annual rainfall varying from 80 inches in the northwestern part of the State to 3 or 4 in the southeastern part, it offers the student of climatology and of the distribution of life facilities unsurpassed in any civilized country, and problems unknown in most other parts of the world.

Long ago geographers divided the earth's surface into five zones, giving them definite boundaries of certain parallels of latitude founded on astronomical considerations. Biologists have also divided the earth's surface into life zones and other divisions. These divisions seldom have very definite boundaries, but blend into one another.

For my present purpose I shall follow the division of the northern hemisphere into three Life Realms, as follows: The Arctic Life Realm, surrounding the north pole and passing southward to the northern limit of trees, or about the annual isotherm of 32 degrees; the North Temperate Life Realm, extending southward from the Arctic Life Realm to about the annual isotherm of 70 degrees; and a Tropical Life Realm. These Life Realms are subdivided into Life Zones as follows: An Arctic Life Zone, consisting of all the Arctic Life Realm; a Boreal Life Zone, consisting of the upper or northern part of the North Temperate Life Realm south to about the summer isotherm of 63 degrees; a Transition Life Zone, consisting of that part of the same Realm bounded above or on the north by the summer isotherm of 63 degrees, and below or south by the summer isotherm of 70 degrees; an Upper Austral Life Zone lying between the summer isotherms of 70 degrees and 77 degrees; a Lower Austral Life Zone, consisting of the remainder of the North Temperate Life Realm and a Sub-Tropical Life Zone, consisting of the northern part of the Tropical Life Realm. This covers but a small area in southeastern California. That part of the Arctic Life Zone in California is still smaller, consisting of a few small isolated areas on the highest mountain summits.

The distribution of life being affected also by the greater or less average amount of moisture present in a given area, and as this average amount of moisture varies in portions of each life zone, it follows that the distribution of life is not equal throughout a life zone. To give expression to the effects of the varying amounts of moisture in life realms and life zones, they are divided in sections of variable size called regions, sub-regions and provinces. That part of the North Temperate Life Realm on this continent is known as the North American Region. That part of this region in western North America having a small annual rainfall is known as the Arid Sub-Region, and the part near the sea having a large rainfall is the Pacific Coast Sub-Region. The Arid Sub-Region has been divided into two provinces: the Sonoran Province, consisting of that part in the Lower Austral and Sub-Tropical Zones; and the Campestrian, consisting of that part in the Upper Austral and Transition Zones.

I propose further subdividing the life areas of California into Faunas, to consist of areas of nearly equal temperature, moisture and soil, and therefore a nearly homogeneous local assemblage of life forms. These will not be equal in either size or value, and are intended only to facilitate the study of distribution of species in California. The boundaries of Life Zones and Faunas as indicated on the accompanying map are only provisional; further study will necessitate numerous changes.

The Californian Arctic Fauna is that part of the Arctic Life

Zone in California. A few species of plants constitute the only peculiarly Arctic life in California, as the areas are so small that animal life of strictly Arctic species has disappeared, with the possible exception of insects.

The Boreal Zone is forested nearly throughout its extent in California. The principal forest trees are the Foxtail Pine, White-barked Pine, Mountain Pine, Tamarack Pine, and Red Fir. The Californian mammals peculiar to this zone are the Gray-headed Pika, Mountain Beaver, Yellow-bellied Marmot, Belding Ground Squirrel, Alpine, Sierra Nevada and Alpine Chipmunks, Californian Pine Squirrel, Black Fox, Wolverine, Pine Marten and Ermine. Some of the birds breeding principally or exclusively in this zone are Sooty Grouse, White-headed Woodpecker, Williamson Woodpecker, Western Nighthawk, Calliope Hummingbird, Olive-sided Flycatcher Gray-eared Finch, White-crowned Sparrow, Lincoln Sparrow, Thick-billed Sparrow, Green-tailed Towhee, Audubon Warbler and Black-throated Gray Warbler. The Californian part of the Boreal Zone may be called the Californian Alpine Fauna.

The Transition Zone is of considerable extent in northern California, but is of less extent in the southern part of the State, where it is limited to the sides and upper parts of the mountains, except that small part rising above about 7,000 feet altitude, which is Boreal. In most parts of the State the Transition Zone is well timbered, and is the great source of supply of wood and lumber in this State. The Yellow, Black and Sugar Pines, White Fir, Cedar and Redwood are characteristic of this zone. It contains a large number of species of birds and mammals, though few, perhaps none, are limited to it, nearly all its species being found in the adjoining zones, either above or below. Some of the birds breeding principally in it are the Californian Woodpecker, Blue-fronted Jay, Californian Purple Finch, Violet-green Swallow and Mountain Chickadee.

The Transition zone in California may be divided into several Faunas. The northeast part of the State, north of Honey

Lake and east of Mt. Shasta, may be called the Modoc Fauna. It is a high broken plateau with some coniferous timber on the highest parts. A character of this Fauna is the abundant presence of sage brush (*Artemesia*). South of the Modoc Fauna is a large area of the Transition Zone in the lower parts of the Sierra Nevada Mountains, which may be called the Sierra Nevada Fauna. It is mostly well timbered, with Yellow Pine as the principal species. Those areas of the Transition Zone lying south of Lat. 35 degrees may appropriately take the name of the San Bernardino Fauna. Here also the Yellow Pine is a characteristic tree. The region about Mt. Shasta, north to Oregon and west to the low strip along the sea coast may provisionally take the name of the Shasta Fauna until its features are better known. I know nothing of this fauna personally, and I can find very little published concerning its faunal conditions. A narrow strip along the seacoast from the Oregon line south to San Francisco may be called the Humboldt Fauna. This is a region of heavy rainfall and fogs, and a strong character is the presence of heavy redwood forests. A continuance of this narrow strip along the coast southward, including the Santa Cruz Mountains, and ending a short distance south of Point Sur, may take the name of the Santa Cruz Fauna. It presents similar characters to that of the Humboldt Fauna, but in a less marked degree.

The Upper Austral Zone lies next below or south of the Transition Zone. In many parts of the Upper Austral Zone a thick growth of several species of shrubs, collectively known as chapparral or chemisal, covers the hills. Forests are few, and west of the Sierras are composed mostly of oaks, which east of the Sierras are replaced by Pinons and Junipers. The Gray-leafed Pine is common in this Zone in some places within the drainage of the San Joaquin and Sacramento valleys. The most characteristic mammals of the Upper Austral Zone are Pocket Rats, two genera and several species, Pocket Mice of several species, Californian Grasshopper Mice, Striped Skunk, Gray and Island Foxes. The following species of birds find their upper or northern limits in this zone: Nuttall Woodpecker, Costa Hummingbird, Yellow-billed Magpie, Nelson Oriole, Lawrence Goldfinch, Black-throated Sparrow, Long-tailed Chat, Californian Thrasher and Black-tailed Gnatcatcher.

That part of the Upper Austral Zone lying on the west side of the Sierra Nevada Mountains, consisting of a long narrow strip along the sides of the lower parts of the mountains, may be called the Foothill Fauna. A broken region of moderate extent, bounded on the west by the Humboldt Fauna, on the north by the Shasta Fauna, on the east and south by the Sacramento Valley, may be called the Clear Lake Fauna. The region bounded on the west and southwest by the Santa Cruz Fauna and the Pacific Ocean, on the southeast by the Santa Ynez Mountains, and on the northeast by the San Joaquin Valley may be called the San Luis Obispo Fauna. All the islands lying off the Southern California coast may be grouped together under the name of the Island Fauna. That part of the Upper Austral Zone south of the San Luis Obispo Fauna and the Mojave Desert and west of the Colorado Desert may be called the San Jacinto Fauna.

The Lower Austral Zone includes most of the Mojave Desert, the San Joaquin and Sacramento valleys, and a strip along the coast from Santa Barbara to San Diego and southward. Over much of this area cactuses form a characteristic part of the vegetation. But few trees occur, and these are found mostly along streams and in damp land. Much of this zone is very arid. Shrews are nearly wanting in this zone. Several species of bats find their northern limit in it, as do several species of ground squirrels. No species of tree squirrels or chipmunks (genera *Sciurus* and *Eutamias*) occur. Several species of Pocket Rats and Pocket Mice and the Big-eared Fox are peculiar to this zone, the Gambel Partridge, Scott Oriole, Leconte Thrasher, Crissal Thrasher, Yellow-headed Tit and Plumbeous Gnatcatcher.

The large valley known as the Sacramento Valley (north-

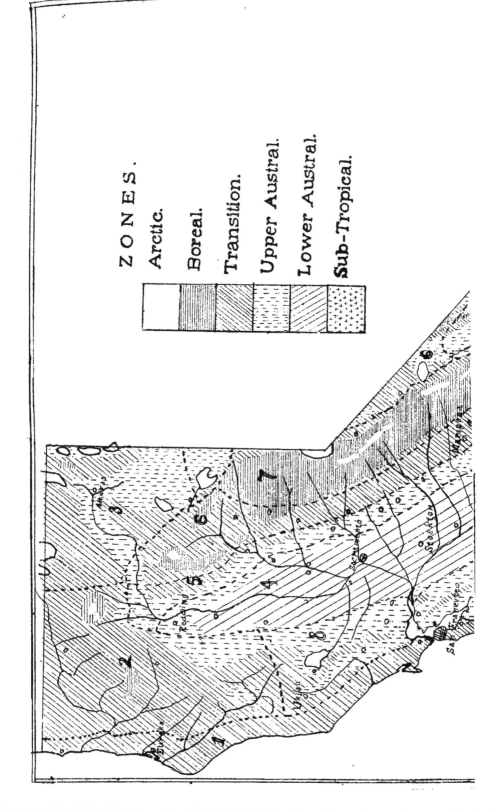

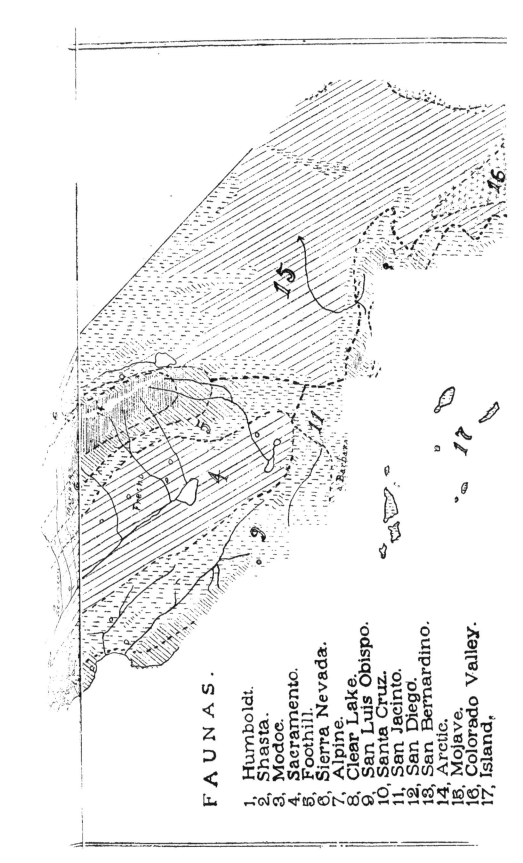

F A U N A S.

1, Humboldt.
2, Shasta.
3, Modoc.
4, Sacramento.
5, Foothill.
6, Sierra Nevada.
7, Alpine.
8, Clear Lake.
9, San Luis Obispo.
10, Santa Cruz.
11, San Jacinto.
12, San Diego.
13, San Bernardino.
14, Arctic.
15, Mojave,
16, Colorado Valley.
17, Island.

ern part), and San Joaquin Valley (southern part), may be called the Sacramento Fauna. The comparatively small area of Lower Austral Zone in the southwestern part of the State may be called the San Diego Fauna. In the eastern part of the State is a large area of arid plain, studded with small barren mountains, known as the Mojave Desert. It is principally Lower Austral Zone, but has a few tracts sufficiently elevated to reach the Upper Austral, and a few very small areas of Transition Zone. This area north of the low Colorado Desert and west of the bottom lands of the Colorado River may be called the Mojave Fauna.

The Sub-Tropical Zone in California is confined to the bottom land along the Colorado River and west in the Colorado Desert, which is properly a part of the same bottom lands. Among the birds which do not breed above this zone, and are found in this part of California are the Harris Hawk, probably the Audubon Caracara, Elf Owl, Vermillion Flycatcher, Abert Towhee and Cooper Tanager. This part of the Sub-Tropical Zone may be called the Colorado Valley Fauna.

List *of*

California Mammals
and their faunal distribution

The Zones are given by name and the Faunas by number. Refer to the map.

Family *BALÆNIDÆ.*

1. *Balæna japonica.* PACIFIC RIGHT WHALE. Pelagic.
2. *Rhachianectes glaucus.* CALIFORNIA GRAY WHALE. Pelagic.
3. *Megaptera nodosa versabilis.* PACIFIC HUMPBACK WHALE. Pelagic.
4. *Balænoptera physalis velifera.* OREGON FINBACK WHALE. Pelagic.
5. *Balænoptera acuto-rostrata davidsoni.* SHARP-HEADED FINNER WHALE. Pelagic.
6. *Sibbaldius sulfureus.* SULPHUR-BOTTLED WHALE. Pelagic.

Family *PHYSETERIDÆ.*

7. *Physeter macrocephalus.* SPERM WHALE. Pelagic.

Family *DELPHINIDÆ.*

8. *Lissodelphis borcalis.* NORTHERN RIGHT WHALE PORPOISE. Pelagic.
9. *Phocæna communis.* BAY PORPOISE. Littoral and pelagic.
10. *Orcinus rectipinna.* STRAIGHT-FINNED KILLER. Pelagic.
11. *Orcinus ater.* BLACK KILLER. Pelagic.
12. *Globicephala scammoni.* SCAMMON BLACKFISH. Pelagic.
13. *Grampus griseus.* COMMON GRAMPUS. Pelagic.
14. *Lagenorhynchus obliquidens.* STRIPED PORPOISE. Pelagic.
15. *Delphinus delphis.* COMMON DOLPHIN. ' Pelagic.
16. *Tursiops gilli.* COWFISH. Pelagic.

Family *CERVIDÆ.*

17. *Cervus roosevelti.* ROOSEVELT WAPITI. Elk.
    Transition, 2.
18. *Cervus nannodes.* CALIFORNIA WAPITI. Elk.
    Lower Austral, 4.
19. *Odocoileus hemionus.* MULE DEER.
    Transition, Boreal, 5, 6, 7.
20. *Odocoileus hemionus eremicus.* BURRO DEER.
    Lower Austral, Sub-tropical, 15, 16.
21. *Odocoileus hemionus californicus.* CALIFORNIA MULE DEER.
    Transition, Upper Austral, Boreal and Lower Austral, 5, 9, 11, 12, 13.
22. *Odocoileus columbianus.* BLACK-TAILED DEER.
    Transition and Boreal, 1, 2, 3, 5, 6.

23. *Odocoileus columbianus scaphiotus.* SOUTHERN BLACK-TAILED DEER.
Transition, Upper Austral, 9, 10.

Family *ANTILOCAPRIDÆ.*

24. *Antilocapra americana.* PRONG-HORNED ANTELOPE.
Upper and Lower Austral, Sub-tropical, 3, 15, 16.

Family *BOVIDÆ.*

25. *Ovis canadensis.* ROCKY MOUNTAIN BIGHORN. Mountain Sheep.
Formerly Boreal, Transition, 2, 3, 6, 7.

26. *Ovis nelsoni.* NELSON BIGHORN...
Transition, Upper and Lower Austral, 13, 15.

27. *Oreamnos montanus.* MOUNTAIN GOAT.
Formerly Boreal, 6, 7.

Family *SCIURIDÆ.*

28. *Marmota flaviventer.* YELLOW-BELLIED MARMOT.
Boreal, 3, 6, 7.

29. *Citellus beecheyi.* CALIFORNIA GROUND-SQUIRREL.
Transition, Upper and Lower Austral, 9, 10, 11, 12, 13.

30. *Citellus beecheyi douglassi.* DOUGLASS GROUND-SQUIRREL.
Transition, Upper Austral, 2, 3, 5, 6.

31. *Citellus beecheyi fisheri.* FISHER GROUND-SQUIRREL.
Upper and lower Austral, 4, 5, 15.

32. *Citellus tereticaudus.* ROUND-TAILED GROUND-SQUIRREL.
Lower Austral, Sub-tropical, 15, 16.

33. *Citellus beldingi.* BELDING GROUND-SQUIRREL.
Boreal, Transition, 3. 7.

34. *Citellus mollis stephensi.* STEPHENS GROUND-SQUIRREL.
Transition, Upper Austral, 6.

35. *Citellus mohavensis.* MOHAVE GROUND-SQUIRREL.
Lower Austral, 15.

36. *Citellus chrysodeirus.* GILDED GROUND-SQUIRREL.
Transition, Boreal, 2, 3, 6, 7.

37. *Citellus chrysodeirus bernardinus.* SAN BERNARDINO GROUND-SQUIRREL.
Transition, Boreal, 13.

38. *Citellus chrysodeirus trinitatus.* TRINITY GROUND-SQUIRREL.
Transition, Boreal, 2.

39. *Citellus leucurus.* ANTELOPE GROUND-SQUIRREL.
Upper and Lower Austral, 3, 6, 15.

40. *Citellus nelsoni.* NELSON GROUND-SQUIRREL.
Lower Austral, 4.

41. *Eutamias alpinus.* ALPINE CHIPMUNK.
Arctic. Boreal, 7, 14.

42. *Eutamias amœnus.* KLAMATH CHIPMUNK.
Boreal, 2, 3, 6, 7.

43. *Eutamias pictus.* DESERT CHIPMUNK.
Upper Austral, Transition, 3, 6.

44. *Eutamias panamintus.* PANAMINT CHIPMUNK.
Transition, Upper Austral, 15.

45. *Eutamias speciosus.* SAN BERNARDINO CHIPMUNK.
Boreal, Transition, 13.

46. *Eutamias speciosus callipeplus.* MOUNT PINOS CHIPMUNK.
Transition, 5, 11.

47. *Eutamias speciosus frater.* SIERRA NEVADA CHIPMUNK.
Transition, Boreal, 6, 7.

48. *Eutamias quadrimaculatus.* LONG-EARED CHIPMUNK.
Transition, Boreal, 6, 7.

49. *Eutamias quadrimaculatus senex.* ALLEN CHIPMUNK.
Transition, Boreal, 3, 6, 7.

50. *Eutamias townsendi ochrogenys.* REDWOOD CHIPMUNK.
Transition, 1.

51. *Eutamias hindsi.* HINDS CHIPMUNK.
Upper Austral, Transition, 1, 8.

52. *Eutamias hindsi pricei.* PRICE CHIPMUNK.
Transition, Upper Austral, 9, 10.

53. *Eutamias merriami.* MERRIAM CHIPMUNK.
Transition, Upper Austral, 5, 11, 13.

54. *Sciurus griseus.* COLUMBIA GRAY SQUIRREL.
Transition, 6, 8.

55. *Sciurus griseus nigripes.* BLACK-FOOTED GRAY SQUIRREL.
Transition, 10.

56. *Sciurus griseus anthonyi.* ANTHONY GRAY SQUIRREL.
Transition, 13.

57. *Sciurus douglassi albolimbatus.* CALIFORNIA CHICKAREE.
Transition, Boreal, 2, 3, 6, 7.

58. *Sciurus douglassi mollipilosus.* REDWOOD CHICKAREE.
Transition, 1.

59. *Sciuropterus alpinus klamathensis.* KLAMATH FLYING-SQUIRREL.
Boreal, 2, 3.

60. *Sciuropterus alpinus californicus.* SAN BERNARDINO FLYING-SQUIRREL.
Boreal, Transition, 13.

61. *Sciuropterus oregonensis stephensi.* STEPHENS FLYING-SQUIRREL.
Transition, 1.

Family *APLODONTIDÆ.*

62. *Aplodontia major.* CALIFORNIA MOUNTAIN BEAVER.
Boreal, 2, 7.

63. *Aplodontia phæa.* POINT REYES MOUNTAIN BEAVER.
Transition, 1.

Family *CASTORIDÆ.*

64. *Castor canadensis frondator.* BROAD-TAILED BEAVER.
Lower Austral, Sub-tropical, 16.
65. *Castor canadensis pacificus.* PACIFIC BEAVER.
Boreal, Transition, Upper Austral, 2, 3, 5, 6.

Family *MURIDÆ.*

66. *Mus norvegicus.* BROWN RAT.
Cosmopolitan.
67. *Mus rattus.* BLACK RAT.
Cosmopolitan.
68 *Mus musculus.* COMMON MOUSE.
Cosmopolitan.
69. *Onychomys torridus ramona.* SAN BERNARDINO GRASSHOPPER-MOUSE.
Upper and Lower Austral, 11, 12.
70. *Onychomys torridus perpallidus.* YUMA GRASSHOPPER-MOUSE.
Sub-tropical, 16.
71. *Onychomys torridus tularensis.* TULARE GRASSHOPPER-MOUSE.
Lower Austral, 4.
72. *Onychomys torridus longicaudus.* LONG-TAILED GRASSHOPPER-MOUSE.
Upper Austral, 15.
73. *Peromyscus texanus gambeli.* GAMBEL MOUSE.
Transition, Boreal, Upper and Lower Austral. Generally distributed.
74. *Peromyscus texanus deserticolus.* DESERT DEER MOUSE.
Lower Austral, Sub-tropical, 15, 16.
75. *Peromyscus texanus clementis.* SAN CLEMENTE MOUSE.
Upper Austral, 17.
76. *Peromyscus oreas rubidus.* MENDOCINO MOUSE.
Transition, 1.
77. *Peromyscus boylii.* BOYLE MOUSE.
Transition, Boreal, Upper Austral, 2, 3, 4, 5, 6, 7, 11.
78. *Peromyscus truei.* BIG-EARED MOUSE.
Upper and Lower Austral, Transition, 1, 2, 8, 9, 10, 11.
79. *Peromyscus californicus.* CALIFORNIA MOUSE.
Upper and Lower Austral, Transition, 1, 5, 8, 9. 10.
80. *Peromyscus californicus insignis.* CHEMISAL MOUSE.
Upper and Lower Austral, 11, 12.
81. *Peromyscus eremicus.* HERMIT MOUSE.
Lower Austral, Sub-tropical, 15, 16.

125. *Thomomys bottæ pallescens.* SOUTHERN POCKET-GOPHER.
Upper and Lower Austral, Transition, 11, 12, 13.

126. *Thomomys laticeps.* BROAD-HEADED POCKET-GOPHER.
Transition, 1.

127. *Thomomys leucodon navus.* RED BLUFF POCKET-GOPHER.
Upper and Lower Austral, 4.

128. *Thomomys angularis.* SAN JOAQUIN POCKET-GOPHER.
Lower Austral, 4.

129. *Thomomys angularis pascalis.* FRESNO POCKET-GOPHER.
Lower Austral, 4.

130. *Thomomys operarius.* OWEN VALLEY POCKET-GOPHER.
Lower Austral, 15.

131. *Thomomys cabezonæ.* CABEZON POCKET-GOPHER.
Lower Austral, 15.

132. *Thomomys fuscus fisheri.* FISHER POCKET-GOPHER.
Transition, 6.

Family *HETEROMYIDÆ.*

133. *Perodipus agilis.* GAMBEL POCKET-RAT.
Upper and Lower Austral, 12.

134. *Perodipus ingens.* BIG POCKET-RAT.
Upper Austral, 9.

135. *Perodipus venustus.* SANTA CRUZ POCKET-RAT.
Transition, 10.

136. *Perodipus goldmani.* GOLDMAN POCKET-RAT.
Upper Austral, 9.

137. *Perodipus panamintus.* PANAMINT POCKET-RAT.
Upper and Lower Austral, 15.

138. *Perodipus streatori.* STREATOR POCKET-RAT.
Upper Austral, 5.

139. *Perodipus microps.* INYO POCKET-RAT.
Lower Austral, 15.

140. *Dipodomys californicus.* CALIFORNIA POCKET-RAT.
Upper Austral, Transition, 1, 8.

141. *Dipodomys californicus pallidulus.* COLUSA POCKET-RAT.
Lower Austral, 4.

142. *Dipodomys deserti.* DESERT POCKET-RAT.
Lower Austral, Sub-tropical, 15, 16.

143. *Dipodomys merriami simiolus.* MIMIC POCKET-RAT.
Lower Austral, Sub-tropical, 15, 16.

144. *Dipodomys merriami parvus.* SAN BERNARDINO POCKET-RAT.
Lower Austral, 12.

145. *Dipodomys merriami nitratus.* KEELER POCKET-RAT.
Lower Austral, 15.

146. *Dipodomys merriami nitratoides.* TULARE POCKET-RAT.
    Lower Austral, 4.

147. *Dipodomys merriami exilis.* LEAST POCKET-RAT.
    Lower Austral, 4.

148. *Microdipodops californicus.* CALIFORNIA DWARF POCKET-RAT.
    Transition, Upper Austral, 3, 6, 7.

149. *Perognathus panamintus.* PANAMINT POCKET-MOUSE.
    Lower Austral, 15.

150. *Perognathus panamintus bangsi.* BANGS POCKET-MOUSE.
    Lower Austral, 15.

151. *Perognathus panamintus arenicola.* SAND POCKET-MOUSE.
    Sub-tropical, 16.

152. *Perognathus brevinasus.* SHORT-NOSED POCKET-MOUSE.
    Lower Austral, 12.

153. *Perognathus pacificus.* SAN DIEGO POCKET-MOUSE.
    Lower Austral, 12.

154. *Perognathus parvus mollipilosus.* COUES POCKET-MOUSE.
    Transition, Boreal, 2, 3.

155. *Perognathus parvus olivaceus.* GREAT BASIN POCKET-MOUSE.
    Upper Austral, Transition, 6.

156. *Perognathus parvus magruderensis.* MT. MAGRUDER POCKET-MOUSE.
    Transition, 15.

157. *Perognathus alticola.* WHITE-EARED POCKET-MOUSE.
    Transition, 13.

158. *Perognathus formosus.* LONG-TAILED POCKET-MOUSE.
    Upper Austral, 15.

159. *Perognathus longimembris.* SAN JOAQUIN POCKET-MOUSE.
    Lower Austral, 4.

160. *Perognathus penicillatus.* TUFT-TAILED POCKET-MOUSE.
    Lower Austral, Sub-tropical, 15, 16.

161. *Perognathus penicillatus angustirostris.* COLORADO DESERT POCKET-MOUSE.
    Sub-tropical, 16.

162. *Perognathus stephensi.* STEPHENS POCKET-MOUSE.
    Lower Austral, 15.

163. *Perognathus fallax.* SHORT-EARED POCKET-MOUSE.
    Upper and Lower Austral, 11, 12.

164. *Perognathus fallax pallidus.* PALLID POCKET-MOUSE.
    Lower Austral, 15.

165. *Perognathus californicus.* CALIFORNIA POCKET-MOUSE.
    Upper Austral, 9.

166. *Perognathus californicus dispar.* ALLEN POCKET-MOUSE.
    Upper and Lower Austral, 5, 9, 11, 12.

167. *Perognathus femoralis.* DARK POCKET-MOUSE.
Upper Austral, 11.
168. *Perognathus spinatus.* SPINY POCKET-MOUSE.
Lower Austral, Sub-tropical, 15, 16.

Family *ZAPODIDÆ.*

169. *Zapus trinotatus.* NORTHWEST JUMPING-MOUSE.
Transition, 1.
179. *Zapus trinotatus alleni.* ALLEN JUMPING-MOUSE.
Boreal, 2, 3, 6, 7.
171. *Zapus orarius.* COAST JUMPING-MOUSE.
Transition, 1.
172. *Zapus pacificus.* PACIFIC JUMPING-MOUSE.
Transition, 2.

Family *ERETHIZONTIDÆ.*

173. *Erethizon epixanthus.* WESTERN PORCUPINE...
Transition, Boreal, 2, 3, 6, 7, 13.

Family *OCHOTONIDÆ.*

174. *Ochotona schisticeps.* SIERRA NEVADA PIKA.
Boreal, 2, 3, 6. 7.

Family *LEPORIDÆ.*

175. *Lepus campestris sierræ.* SIERRA PRAIRIE HARE.
Boreal, Transition, 7.
176. *Lepus californicus.* CALIFORNIA HARE.
Upper and Lower Austral, Transition. 2, 4, 5, 8, 9, 11, 12.
177. *Lepus richardsoni.* RICHARDSON HARE.
Upper Austral, 5, 9.
178. *Lepus texianus deserticola.* DESERT HARE.
Upper and Lower Austral, Sub-tropical, 3, 6, 15, 16.
179. *Lepus texianus tularensis.* TULARE HARE.
Lower Austral, 4.
180. *Lepus auduboni.* AUDUBON HARE.
Upper and Lower Austral, Transition, 4, 5, 6, 8, 10, 11, 12, 13.
181. *Lepus auduboni arizonæ.* ARIZONA WOOD HARE.
Upper and Lower Austral, Sub-tropical, 15, 16.
182. *Lepus nuttalli.* NUTTALL WOOD HARE.
Transition, 2, 3.
183. *Lepus bachmani.* BACHMAN BRUSH HARE.
Transition, 1, 10.
184. *Lepus cinerascens.* ASHY BRUSH HARE.
Upper and Lower Austral, 9, 11, 12.

Family *PHOCIDÆ*.

185. *Phoca richardii.* PACIFIC HARBOR SEAL.
Littoral.

186. *Phoca richardii geronimensis.* SAN GERONIMO HARBOR SEAL.
Littoral.

187. *Mirounga angustirostris.* CALIFORNIA ELEPHANT SEAL.
Littoral.

188. *Zalophus californianus.* CALIFORNIA SEA LION.
Littoral.

189. *Eumetopias jubata.* STELLAR SEA LION.
Littoral.

190. *Callorhinus alascanus.* NORTHERN FUR SEAL.
Littoral.

191. *Arctocephalus townsendi.* GUADALOUPE FUR SEAL.
Littoral.

Family *FELIDÆ*.

192. *Felis hippolestes olympus.* PACIFIC COAST COUGAR.
Transition, Upper Austral, 1, 2, 3, 5, 6, 8.

193. *Felis aztecus browni.* BROWN COUGAR.
Transition, Upper and Lower Austral, Sub-tropical, 11, 12, 13, 15, 16.

194. *Lynx eremicus.* DESERT LYNX.
Upper and Lower Austral, Sub-tropical, 5, 8, 9, 11, 12, 13, 15, 16.

195. *Lynx fasciatus pallescens.* WASHINGTON LYNX.
Transition, Upper Austral, 1, 2, 3, 5, 6.

Family *CANIDÆ*.

196. *Canis ochropus.* VALLEY COYOTE.
Upper and Lower Austral, Transition. 2, 4, 5, 8, 9, 11, 12.

197. *Canis estor.* DESERT COYOTE.
Upper and Lower Austral, Sub-tropical, 15, 16.

198. *Canis lestes.* MOUNTAIN COYOTE.
Boreal, Transition, 2. 3, 6, 7.

199. *Canis mearnsi.* MEARNS COYOTE.
Transition, 11. 13.

200. *Canis mexicanus.* GRAY WOLF.
Boreal, Transition, 3, 7.

201. *Vulpes macrotis.* LONG-EARED FOX.
Lower Austral, Sub-tropical, 11, 15, 16.

202. *Vulpes muticus.* SAN JOAQUIN FOX.
Lower Austral, 4.

203. *Vulpes necator.* HIGH SIERRA FOX.
Boreal, 7.

204. *Vulpes cascadensis.* CASCADE MOUNTAIN FOX.
    Boreal, 2, 3, 7.

205. *Urocyon californicus.* CALIFORNIA GRAY FOX.
    Transition, Boreal, Upper and Lower Austral, 9, 11, 12, 13.

206. *Urocyon californicus townsendi.* TOWNSEND GRAY FOX.
    Transition, Upper Austral, 2, 3, 5, 6.

207. *Urocyon littoralis.* SAN MIGUEL ISLAND FOX.
    Upper Austral, 17.

208. *Urocyon littoralis santacruzæ.* SANTA CRUZ ISLAND FOX.
    Upper Austral, 17.

209. *Urocyon clementæ.* SAN CLEMENTE ISLAND FOX.
    Upper Austral, 17.

210. *Urocyon catalinæ.* SANTA CATALINA ISLAND FOX.
    Upper Austral, 17.

Family *PROCYONIDÆ.*

211. *Bassariscus astutus raptor.* CALIFORNIA RING-TAILED CAT.
    Transition, Upper Austral, 1, 2, 3, 5, 6, 13.

212. *Procyon psora.* CALIFORNIA RACCOON.
    Upper and Lower Austral. Transition, 4, 5, 8, 11, 12.

213. *Procyon psora pacifica.* PACIFIC RACCOON.
    Upper Austral, Transition, 2.

214. *Procyon pallidus.* DESERT RACCOON.
    Lower Austral, Sub-tropical, 15, 16.

Family *URSIDÆ.*

215. *Ursus horribilis.* GRIZZLY BEAR.
    Transition, Upper and Lower Austral, 2, 3, 4, 5, 6, 8, 9, 13.

216. *Ursus americanus.* BLACK BEAR.
    Boreal, Transition, Upper Austral, 1, 2, 3, 5, 6, 8, 9, 13.

Family *MUSTELIDÆ.*

217. *Latax lutris nereis.* SOUTHERN SEA OTTER.
    Littoral.

218. *Lutra canadensis pacifica.* PACIFIC OTTER.
    Transition, Upper Austral, 1, 2, 3, 4, 5, 6, 8.

219. *Lutra canadensis sonora.* SONORA OTTER.
    Lower Austral, Sub-tropical, 16.

220. *Taxidea taxus neglecta.* WESTERN BADGER.
    Transition, Upper and Lower Austral, 2, 3, 4, 5, 6, 8, 9, 11, 12.

221. *Mephitis occidentalis.* CALIFORNIA SKUNK.
    Transition, Upper and Lower Austral, 1, 2, 4, 5, 8, 9, 10.

222. *Mephitis occidentalis major.* GREAT BASIN SKUNK.
    Transition, Upper Austral, 3, 6.

223. *Mephitis occidentalis holzneri.* SOUTHERN CALIFORNIA SKUNK.
Transition, Upper and Lower Austral, 9, 11, 12, 13.

224. *Mephitis platyrhinus.* BROAD-NOSED SKUNK.
Transition, Upper Austral, 5, 6.

225. *Mephitis estor.* ARIZONA SKUNK.
Lower Austral, Sub-tropical, 15, 16.

226. *Spilogale phenax.* WESTERN SPOTTED SKUNK.
Upper and Lower Austral, 4, 5, 9, 11, 12.

227. *Spilogale latifrons.* LITTLE SPOTTED SKUNK.
Upper Austral, Transition, 2.

228. *Gulo luscus.* WOLVERINE.
Boreal, Transition, 2, 3, 6, 7.

229. *Mustela pennanti pacifica.* PACIFIC FISHER.
Boreal, Transition, 1, 2, 3, 6, 7.

230. *Mustela caurina.* PACIFIC PINE MARTEN.
Boreal, Transition, 1, 2, 3, 6, 7.

231. *Lutreola vison energumnos.* PACIFIC MINK.
Transition, 1, 2, 3, 6, 7.

232. *Putorius xanthogenys.* CALIFORNIA WEASEL.
Upper and Lower Austral, Transition, 4, 5, 9, 11, 12.

233. *Putorius xanthogenys mundus.* REDWOODS WEASEL.
Transition, Upper Austral, 1, 8.

234. *Putorius arizonensis.* MOUNTAIN WEASEL.
Transition, Boreal, 2, 3, 6, 7.

235. *Putorius muricus.* LITTLE WEASEL.
Boreal, 7.

Family *SORECIDÆ.*

236. *Sorex vagrans.* WANDERING SHREW.
Transition, Boreal, Upper Austral, 1, 2, 3, 6, 7, 9, 10, 11.

237. *Sorex amœnus.* SIERRA NEVADA SHREW.
Boreal, 7.

238. *Sorex obscurus.* DUSKY SHREW.
Boreal, 7.

239. *Sorex montereyensis.* MONTEREY SHREW.
Transition, Boreal, 1, 2, 5, 6, 9, 10.

240. *Sorex ornatus.* ADORNED SHREW.
Transition, Upper Austral, 11, 13.

241. *Sorex californicus.* CALIFORNIA SHREW.
Transition, Upper Austral, 8, 9.

242. *Sorex tenellus.* INYO SHREW.
Transition, 6.

243. *Sorex tenellus lyelli.* MOUNT LYELL SHREW.
Boreal, 7.

244. *Sorex tenellus myops.* WHITE MOUNTAIN SHREW.
    Transition, 6.
245. *Sorex pacificus.* PACIFIC SHREW.
    Transition, 1.
246. *Sorex palustris navigator.* WATER SHREW.
    Boreal, Transition, 2, 3, 7.
247. *Sorex bendirei.* BENDIRE SHREW.
    Transition. 1.
248. *Notiosorex crawfordi.* GRAY SHREW.
    Upper and Lower Austral, 11, 12.

Family *TALPIDÆ.*

249. *Scapanus townsendi.* TOWNSEND MOLE.
    Transition, 1.
250. *Scapanus californicus.* CALIFORNIA MOLE.
    Transition, Upper Austral, 1, 2, 5, 6, 8, 10.
251. *Scapanus californicus anthonyi.* ANTHONY MOLE.
    Transition, Upper and Lower Austral, 11, 12, 13.
252. *Scapanus californicus truei.* MODOC MOLE.
    Upper Austral, 3.
253. *Neurotrichus gibbsi major.* LARGE SHREW-MOLE.
    Transition, Boreal, 2, 3.

Family *VESPERTILIONIDÆ.*

254. *Antrozous pallidus.* PALE BAT.
    Lower Austral, Sub-tropical, 15, 16.
255. *Antrozous pallidus pacificus.* PACIFIC PALE BAT.
    Upper and Lower Austral, 4, 5, 9, 11, 12.
256. *Euderma maculatum.* SPOTTED BAT.
    Lower Austral, Sub-tropical, 12, 16.
257. *Corynorhinus macrotis pallescens.* LUMP-NOSED BAT.
    Lower Austral, 12, 15.
258. *Myotis lucifugus longicrus.* LONG-SHANKED BAT.
    Transition, Upper Austral, 2, 3, 5, 11, 15.
259. *Myotis californicus.* LITTLE CALIFORNIA BAT.
    Upper and Lower Austral, Transiton, 1. 2, 5, 8, 9, 10, 11, 12, 15, 17.
260. *Myotis californicus pallidus.* PALLID BAT.
    Lower Austral, Sub-tropical, 15, 16.
261. *Myotis yumanensis.* YUMA BAT.
    Upper and Lower Austral, Sub-tropical, 3, 4, 5, 11, 12. 15, 16.
262. *Myotis yumanensis saturatus...* MILLER BAT.
    Boreal, Transition, 2, 3.
263. *Myotis evotis.* LONG-EARED BAT.
    Transition, Upper and Lower Austral, 2, 3, 6, 11, 12. 15.

264. *Myotis thysanodes.* FRINGED BAT.
Upper Austral, 11.

265. *Lasionycteris noctivagaus.* SILVERY-HAIRED BAT.
Transition, 1, 5, 8.

266. *Pipisterllus hesperus.* WESTERN BAT.
Upper and Lower Austral, Sub-tropical 5, 11, 12, 15, 16.

267. *Eptesicus fucus bernardinus.* SAN BERNARDINO BAT.
Transition, Boreal, Upper Austral, 9, 11, 12, 13.

268. *Eptesicus fuscus melanopterus.* SIERRA BAT.
Transition, Boreal, 1, 2, 3, 5, 6, 7, 8.

269. *Lasiurus borealis teliotis.* WESTERN RED BAT.
Upper and Lower Austral, 4. 9, 11, 12.

270. *Lasiurus cinereus.* HOARY BAT.
Transition, Upper and Lower Austral, 4, 9, 11, 12.

Family *MOLLOSSIDÆ.*

271. *Nyctinomops mohavensis.* MOHAVE BAT.
Upper and Lower Austral, Sub-tropical, 9, 12, 15, 16.

272. *Nyctinomops femorosaccus.* POCKETED BAT.
Sub-tropical, 16.

273. *Nyctinomops depressus.* NEVADA BAT.
Lower Austral, 15.

274. *Promops californicus.* CALIFORNIA MASTIFF BAT.
Lower Austral, 12.

Family *PHYLLASTOMATIDÆ.*

275. *Otopterus californicus.* CALIFORNIA LEAF-NOSED BAT.
Upper and Lower Austral, Sub-tropical, 11, 12, 15, 16.

Family *HOMINIDÆ.*

276. *Homo sapiens americanus.* AMERICAN INDIAN. Throughout California.

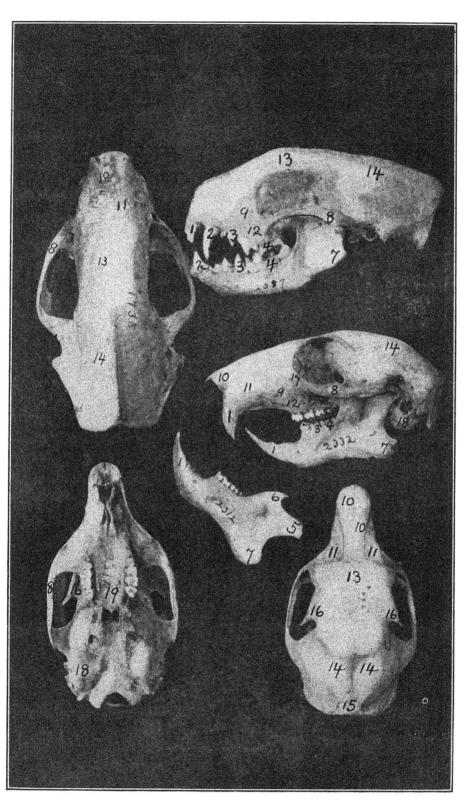

Parts of a Skull

A skull is composed of a number of bones. The size and shape of these bones vary more or less with different species, and they therefore form good characters for distinguishing species and groups. The sutures indicating the lines of junction between adjacent bones anchylose more or less in aged animals and become obscure. Below are the names of the principal parts of a skull, with figures corresponding to those on the plate opposite.

Skulls Nos. 1738 and 2087 are Southern California Skunks. Nos. 696, 2312 and 2332 are Columbia Gray Squirrels.

1. Incisor.
2. Canine.
3. Premolar.
4. Molar.
5. Condyle.
6. Coronal Process.
7. Angular Process.
8. Zygomatic Arch.
9. Anteorbital Foramen. (The figure shows the location the orifice.)
10. Nasal.
11. Premaxillary.
12. Maxillary.
13. Frontal.
14. Parietal.
15. Interparietal.
16. Postorbital Process.
17. Malar.
18. Audital Bulla.
19. Palate.

Glossary

Abnormal. Irregular. Differing from the usual character.

Adult. Full grown.

Affinity. Direct relationship.

Alpine. Used here as pertaining to high altitudes, chiefly near timber line.

Analogy. Superficial resemblance without direct relationship.

Animal. A living (animated) creature, capable of growth and voluntary motion. Often, but wrongly, this term is restricted to mammals.

Aquatic. Pertaining to, or living in the water.

Arboreal. Pertaining to, or living in trees.

Bicolor. Of two colors.

Biology. The study of all living things.

Boreal. Northern.

Canine. The conical tooth next the incisors; wanting in rodents.

Carnivorous. Flesh-eating.

Character. Any peculiarity available for diagnosis.

Chemisal. Thickets of chemise and other brush, such as cover the hillsides of Southern Californa.

Classification. A systematic arrangement.

Clavicle. The collar bone.

Congeneric. Of the same genus.

Crepuscular. Active at twilight.

Deciduous. Shed at certain periods.

Diagnosis. In taxonomy, a condensed statement of a set of characters applicable to a group of animals.

Dichromatic. Having two phases of color independent of age, sex or season.

Digitigrade. Walking on the toes. Includes all birds and most mammals.

Diurnal. Active in the daytime.

Dorsal. Pertaining to the back.

Embryo. Earlier stages of unhatched or unborn young.

Exotic. Foreign.

Family. A group of genera agreeing in certain characters and differing from other families of the order in one or more characters.

Fauna. The animal life of a region.

Fissiped. Having cleft toes.

Fluvatile. Inhabiting rivers.

Fœtus. Later stages of unborn young.

Foramen. A hole or opening, usually small.

Fossorial. Inhabiting burrows.

Frugivorous. Fruit eating.

Genus. A group of species agreeing in certain characters and differing from other genera of the family in one or more characters.

Gregarious. Going in flocks or herds.

Habitat. Region inhabited by a species.

Heterogenous. Of unlike or miscellaneous characters.

Hibernate. To become torpid in winter quarters.

Homogenous. Of like characters.

Humerus. The bone of the upper part of the fore limb, from shoulder to elbow.

Hybrid. Progeny of parents of different species.

Immature. Not mature; ungrown.

Incisor. Tooth in the front part of mouth, between the canines.

Indigenous. Native in a region.

Insectivorous. Feeding on insects.

Interfemoral membrane. The membrane connecting the hind legs of a bat.

Littoral. Pertaining to the shore.

Longitudinal. Lengthwise.

Maculate. Spotted.

Mammal. Animals that suckle their young.

Manus. The hand or fore foot.

Marine. Of the sea.

Maratime. Pertaining to the border of the sea.

Molar. A grinding tooth.

Monogamous. Mating with a single individual of the opposite sex.

Nocturnal. Active in the night.

Normal. Of the usual character; standard; regular.

Omnivorous. Feeding on anything eatable.

Order. A group of families agreeing in certain characters.

Pelage. A covering of hair.

Pelagic. Frequenting the sea far from land.

Pes. The hind foot.

Piscivorous. Feeding on fish.

Plantigrade. Walking on the full length of the foot.

Polygamous. Mating with more than one female.

Premolar. The anterior permanent molariform teeth which are preceded by deciduous ("milk") teeth.

Radius. The front one of the two bones of the forearm.

Retractile. Capable of being drawn back, like a cats claw.

Rostrum. The beak; the front part of the skull.

Sectorial. Adapted for cutting.

Septum. A partition.

Species. An assemblage of related individuals agreeing in certain characters but differing distinctly in one or more characters (usually of minor importance) from all other individuals of the same condition.

Subspecies. A form connected by other forms of the species by intermediate individuals; a variety; a nascent species; a geographical race.

Subterranean. Under the surface of the earth; underground.

Synonym. A duplicate name discarded for a prior name, or one more applicable.

Tarsus. The shank bone; the bones between the toes and the heel.

Taxidermist. One who prepares skins to imitate the form of living animals.

Taxonomy. The science of classification.

Terrestrial. On the surface of the ground.

Tibia. The larger bone below the knee.

Tragus. The inner lobe of the ear.

Type. Of a species, that specimen used as a base of the description of the species in the original description; of a genus, that species from which the generic characters were taken.

Ulna. The back one of the two bones of the forearm; that one on which the elbow hinges.

Vertebrate. Having a spinal column or backbone.

Zoology. The natural history of animals in general.

Errata

Page 22, next to bottom line, for *havits* read *habits*.

40. Tenth line from top, for *clusely*, read *closely*.

116. Third line from bottom, for *marcotis* read *macrotis*.

162. For Panamint *Pocket-Rat,* read Panamint *Pocket-Mouse.*

114. For Subgenus *Neotome* read *Teomoma.*

195. Head, for *Feræ* read *Leporidæ.*

241. Head, for Mustedidæ, read Mustelidæ.

Index